Empire and the Sun

WRITING SCIENCE

EDITORS Timothy Lenoir and Hans Ulrich Gumbrecht

EMPIRE AND THE SUN

VICTORIAN SOLAR ECLIPSE EXPEDITIONS

Alex Soojung-Kim Pang

Stanford University Press
Stanford, California

Stanford University Press
Stanford, California
© 2002 by the Board of Trustees of the
Leland Stanford Junior University
Printed in the United States of America

Library of Congress Cataloging-in-Publication Data

Pang, Alex Soojung-Kim
    Empire and the sun : Victorian solar eclipse expeditions / Alex Soojung-
Kim Pang.
        p.   cm. — (Writing science)
    Includes bibliographical references and index.
    ISBN 0-8047-3925-0 (acid-free paper) —
    ISBN 0-8047-3926-9 (pbk. ; acid-free paper)
    1. Solar eclipses—Observations—History—19th century.
    2. Scientific expeditions—Great Britain—History—19th century.
    I. Title. II. Series.

QB541.P36   2002
523.7'8'09034—dc21   2001057661

This book is printed on acid-free, archival-quality paper.

Original printing 2002
Last figure below indicates year of this printing:
    11   10   09   08   07   06   05   04   03   02

Typeset at Stanford University Press in 10/13 Sabon

To Rob and Riki

Acknowledgments

In keeping with the spirit of this book, my acknowledgments begin in the field. All historians owe debts to archivists and librarians, and when working in odd areas and unusual subjects—such as eclipse expeditions—one relies heavily on archivists' intimate familiarity with their collections. Therefore, I am grateful for the help of Peter Hingley and Mary Chibnall at the Royal Astronomical Society; Adam Perkins at the Royal Greenwich Observatory Archives, Cambridge University; and Angus MacDonald and his staff at the Royal Observatory at Edinburgh. I also thank the reference and interlibrary loan staffs at the University of Pennsylvania, Williams College (especially Karen Worley), Stanford University (in particular Sonia Moss), and the University of California, Berkeley.

My advisors at the University of Pennsylvania, Rob Kohler and Riki Kuklick, helped me develop this project into a dissertation, and continued to provide helpful feedback through later revisions. The Benjamin Franklin Scholars and University Scholars programs gave me an essential head start in my scholarly pursuits, in the form of opportunities unavailable to most undergraduates to study with exceptional faculty and conduct research. The Gaius Bolin Minority Predoctoral Fellowship at Williams College provided support for archival work, and an excellent environment in which to begin writing. My fellow graduate students in the Department of History and Sociology of Science—especially Jeff Brosco and Amy Slaton—were an affectionate, stimulating group, to whom I owe more than I can say.

The bulk of the rewriting was done on an NSF postdoctoral fellowship at Stanford University, and a Chancellor's Minority postdoctoral fellowship at the University of California, Berkeley. At Stanford, I was lucky to share the valuable company of fellow postdocs Betty Smocovitis, Thomas Soderqvist, and Michael Dettelbach. The generosity of the Chancellor's Fellowship at the University of California, Berkeley, allowed me the luxury of immersing myself in the project for two years under nearly ideal conditions. Berkeley's public culture of cafes and bookstores provided an ideal atmosphere for academic pursuits; in particular, Cafe Milano, where most of the actual writing was done, became my home away from home. Most important, my friends—especially Sonja Amadae, Alexi Assmus, David Hollinger, Christopher Kutz, My Le, Abigail Lustig, Nick McKeown, Jessica Riskin, and Susan Spath—read drafts of chapters, listened to talks, listened to me talk about the project, and gave moral support.

Earlier versions of Chapters 3 and 4 appeared in *Isis* and the *Journal for the History of Astronomy*. Other sections were improved by feedback from audiences at RPI, U.C. Berkeley, U.C. San Francisco, and the 1989 and 1992 History of Science Society conferences. Many of the ideas in Chapter 5 developed out of my seminar on science and imperialism, and I thank my students, especially John Bartol and Anne Fadiman, for many stimulating discussions.

Finally, I thank my parents, who taught me much about why history matters, my wife, Heather, who never let me forget the manuscript, and my daughter Elizabeth, who likes to play on the keyboard.

Contents

(12 pages of photographs follow page 82)

Tables

Empire and the Sun

CHAPTER I

Introduction

Why Eclipse Expeditions?

This book is about the practice and experience of science in the Victorian era. It focuses on solar eclipse expeditions, in which astronomers carried telescopes and spectroscopes to remote areas of India, the coasts of Africa, the mountains of Japan, the Great Plains of North America, and islands in the Caribbean and the Pacific, to watch the sun eclipsed by the moon for as long as seven minutes. Eclipse expeditions can be of considerable scientific importance, but they're quite different from run-of-the-mill astronomy, which is usually done not in the field but in the observatory, at the blackboard, or more recently at the PC or workstation. Still, some Victorian astronomers loved them, and my first task is to explain why a reader might share that enthusiasm. In opening this way I follow a long tradition, for books on eclipses have often started with an apologia. Almost exactly a century ago, the Rev. William Lynn felt obliged to defend his book *Remarkable eclipses:* "Perhaps," he imagined, "someone will exclaim, 'What, another book on eclipses? Are there not enough already?' The author can only reply, 'Is this not a little one?' "[1]

There are fewer books written about eclipses these days, and this one is not so little. It is my aim in this chapter to explain what historians of science can learn from this topic. Certainly other sorts of expeditions received more notice in their day, and to the modern reader the exploits of geographers like Sir Richard Burton, or naturalists like Charles Darwin and Henry Walter Bates, are better known than those of any characters in my story. Compared to these expeditions, which lasted years and whose impact was measured over lifetimes (if they did not claim lives, as so often happened), eclipse expeditions seem quite tame. Astronomers put up with sleepless nights, anxiety, trouble with instruments, problems with assistants, nightmares with servants, and difficulties getting good food in the field, but few encounters with dangerous animals and disease, and none with death.

TABLE I

Eclipses Observed by British Parties, 1842–1905

Year	Area visible	Organizers
1842	Scandinavia	Private
1851	Europe	Private
1860	Spain	George Airy et al.
1868	India	James Tennant (w/ Airy et al.)
1870	Spain and Mediterranean	RAS-Royal Society
1871	India, Ceylon, Australia	BAAS
1875	Far East	RAS
1878	United States	Private (Norman Lockyer)
1882	Egypt	Private (Norman Lockyer)
1883	South Pacific	Private (Norman Lockyer)
1886	Caribbean	RAS-Royal Society
1889	South America, West Africa	Royal Society
1893	West Africa	RAS-Royal Society-Solar Physics Committee
1896	Scandinavia, Russia, Japan	JPEC, BAA
1898	India	JPEC, BAA
1900	United States, Southern Europe	JPEC, BAA
1901	Sumatra	JPEC, BAA
1905	Spain, Egypt	JPEC, BAA

But these events still have the power to draw a historian's attention, even if they are not so well known or represent an extreme of human experience. Eclipse expeditions produced mountains of documentation and artifacts, making it possible to reconstruct the planning of expeditions, life in the field, observational practices, and the emotional texture of fieldwork in considerable detail. Preliminary proposals, research agendas, minutes of meetings, correspondence with funding agencies and foreign governments, observing notes, letters from the field, diaries, transcripts of speeches, drafts of reports, and accounts published in newspapers, popular magazines, scientific journals, books, and autobiographies; drawings of eclipses; and photographs of camps and instruments, scientists and assistants at work, the solar corona and the space around the sun—these materials fill file after file in archives, recording in detail every aspect of expeditionary work and life. One reason expeditions were so heavily documented was that within the world of Victorian astronomy they were a big deal. Astronomy was a popular and important part of Victorian science, so this was a pretty big world, but it was also a highly organized one. By watching it choose observers, equip parties, and send expeditions out into the field, we can learn things about its structure and behavior, and improve our understanding of the organization and politics of Victorian science more generally.

Eclipse expeditions are also interesting for the ways they reflect Victorian imperial culture and values, and expand our view of the Victorian field sciences. Planners used eclipses to link astronomy to politics and science to imperialism, and our current interest in the history of such connections should make us keenly attentive to the ways they were articulated in earlier times. But reports from the field, narratives of expeditions, and memoirs also lay bare deeper connections between science and colonial culture that either went unremarked or were so commonly known as to be overlooked. The fact that eclipse fieldwork was so different from the grand expeditions of geographers or naturalists—in their duration, size, funding—also is a real advantage, for it broadens our understanding of what scientific fieldwork is like, and throws some new light on those kinds of better-known expeditions.

Most of my story deals with expeditions officially sponsored by major British scientific societies, the Royal Astronomical Society, Royal Society of London, and British Association for the Advancement of Science. Occasionally I will mention amateur scientists who traveled independently, but I spend little time on them for several reasons. Officially sponsored expeditions were more important scientifically. They included more serious and advanced observers, and had access to better instruments, more money, official favors, and all the other things that make good science easier. Sources also dictate this focus: official expeditions left records in half a dozen archives, while independent expeditions are documented in a handful of published memoirs. For similar reasons, I have little to say about the role of women on expeditions. Women were not major players in eclipse fieldwork, but they did appear on enough expeditions to deserve some discussion; however, British sources are notable for their steadfast silence on the roles women played in the field.[2] Similarly, a couple of diary entries and brief recollections in memoirs are all that document the experiences of natives, sailors, or soldiers who were enlisted as assistants or guards. As for indigenous peoples, the subjects of so much derision and fear on the part of British astronomers, there are no sources that describe the experience of eclipse-watching from their point of view. This is especially unfortunate, for a comparison of the experiences of watching an eclipse—and of watching members of another culture watch an eclipse—would have been interesting. As it is, we can look only one way, from European to native. Finally, I have spent very little time discussing spectroscopic observation or theories of the sun's constitution, as these subjects have been dealt with very well by other authors.[3] Instead, I have chosen to concentrate on visual representation, and the strategies and technologies astronomers developed to produce drawings and photographs of eclipses.

Historiographic Context

Three scholarly literatures constitute the immediate background to this book: the diverse body of work referred to as science and technology studies (STS); the no less varied literature on visual representation and science; and the more focused and unified body of work on science and empire.

SCIENCE AND TECHNOLOGY STUDIES

Science and technology studies is a loose confederation of books and articles unified by the contention that scientific research and discovery is a process of "social construction" influenced as much by social and cultural forces—traditionally considered "external" to science—as by evidence, logic, and theories—factors traditionally considered "internal" to science.[4] One strand of this literature consists of ethnographic studies of scientific work, which set out to show that the everyday world of the laboratory is quite different from that postulated by philosophers. Those differences are often most dramatically brought forth by scientific controversies, during which participants articulate usually tacit assumptions, wrangle over methodological issues, and fight over the rules by which arguments will be settled.[5] A second strand, represented most famously by members of the "Edinburgh School," developed historical and contemporary studies informed by the work of Thomas Kuhn, Robert Merton, and other philosophers or sociologists. Their most notable achievement has been to argue that social, economic, and cultural "interests" play a role in the shaping of both good and bad science. In the mid-1980s, Steven Shapin and Simon Schaffer's *Leviathan and the air-pump* and Martin Rudwick's *The great Devonian controversy* announced a synthesis of these several strands, and the emergence of a new style of scholarship in science studies. In the years since, the field has continued to diversify, as new scholarly tools are applied and subjects opened up for investigation. Perhaps no development has been more influential than the emergence of "practice" as a critical research area, and the reconstruction in elaborate detail of the way scientists work as a key technique for understanding the construction of science.[6] These studies often emphasize the material aspects of scientific work over the social or cognitive. Laboratory animals and workplaces have both received attention on these grounds, but both have been outstripped by instruments—the subject of so much attention by scientists, the producers of scientific data, the ground upon which theory and experiment meet. Studies of practice have also focused on people such as technicians, assistants, artists, and administrators, figures who are essential members of the research enterprise but normally get pushed into the background.[7] Another group has applied the tools of cultural studies and literary theory to science, arguing (or concluding) that local culture and discourse are of overriding importance in explaining the development of scientific ideas. More recently, among scholars raised in the constructivist tradition, the overtones of the term "social construction" have changed in yet a different way. They have devoted more attention to the "construction" side of the phrase, showing how facts are manufactured and negotiated through practices and technologies, rather than defined by interests and ideologies.[8] This work takes for granted what was debated in the 1970s and 1980s—that social interests help shape scientific ideas—and sets out instead to show how social construction works: how people work through epistemological uncertainties, technical difficulties, and theoretical conundrums to produce knowledge that seems trustworthy. It also pays greater attention to the constraints imposed by nature on scientific practice and inquiry, though the meaning and use of the term "constraint" is still a matter of debate.[9]

This book is part of the emerging "postconstructivist" strain in science studies.[10] Its desire to recapture the emotional texture of science along with the messy details of its practice further sets it apart from traditional social constructivism. In their quest to document the effects of social interests on science, the Edinburgh School and its allies put aside the task of reconstructing the emotional texture of scientific practice and discovery. Perhaps, in order to establish the value of the externalist program, it seemed as necessary to move past journalistic accounts that turned science into an adventure as it was to move past philosophical accounts that reduced it to an exercise in logic. But replacing the narrative voice with a more rigorously analytical one has disturbed some critics, who protest that "externalist" or social histories of science downplay the difficulty with which new facts about nature are made. Some have gone even further and argue that constructivist accounts of science make forces like ideology and power, which are more normally thought of as distorting science rather than supporting it, more central than the ordinary work that scientists do in the laboratory or field.[11] The social constructivist program's notable successes in linking ideas about nature to social order and ideological commitments have been bought by sacrificing a sense of what it feels like to build a camp, handle a new instrument in the observatory, deal with solitude and worry in the field, master a difficult technique, make a discovery, accept a failure. Carefully crafted microstudies have generally not tried to reconstruct the emotional or psychological dimension of science for its practitioners. My own feelings about doing research has convinced me that we ignore this aspect of science at our own peril. The irrational exuberance I feel in libraries and archives, the satisfaction of resisting temptations and diversions to work, the agony and pleasure I get while writing, and the pleasure I enjoy after solving a knotty historiographic problem all explain why I became a scholar, and why I continue to enjoy the life of the mind in the face of mixed professional rewards. Doing history is an addiction, not just a job. I'm certain that some of my subjects would have understood my feelings.

Passions, like social interests, are hard to document and very difficult to link to intellectual commitments or reference for one or another instrument or technique. Popular accounts of science that play up eureka moments, the excitement of the experimental chase, and the thrill of discovery show that this can be done badly. But as the work of Carlo Ginzburg, Joan Scott, and others suggests, an analysis of experience and *mentalites* might reveal links between science and other aspects of practitioners' worlds. The effort to uncover the links between social allegiances, politics, and scientific ideas required downplaying psychological and emotional considerations. Now that the place of "external" factors in shaping science has been established, it is time to bring back into our accounts the feeling and flavor of doing science. The excitement, passion, and worry that scientists feel about their work should also be taken seriously, because they offer clues about the organization of work and the relationship between identity and scientific practice. By replacing "actants" with living and feeling people, we can discover links between social life, culture, and work that otherwise go unrecorded, and build a richer and more complete picture of scientific practice.[12]

VISUAL REPRESENTATION IN SCIENCE

The literature on visual representation in science also influences this book. The subject of visual representation has enjoyed a tremendous upsurge in popularity in the last decade. Its emergence follows the turn to the study of instruments and practices in science. Instruments are often designed to generate visual records, while the study of scientific practice underwrote a broad view of scientific work that legitimated analysis of a wide range of visual materials produced by a variety of practitioners.[13] Visual records have also been appealing because they seem well suited to revealing the constructed, contingent aspect of scientific knowledge. Visual representations are never natural objects; they are "ordered, shaped, and filtered . . . a rich repository [and target] of social actions" expressing a relationship between an instrument or observer and nature, between instruments, or between other representations. The process of making pictures involves "an active reconstruction of the world" in which social interests can be seen at work.[14]

The importance of imaging techniques and technologies to a history of eclipse expeditions is obvious. One of the great objectives of eclipse observers was to record the appearance of the solar corona, and that had to be done amid the many distractions thrown up before and during an eclipse. Much of my analysis follows the conventions of the literature on visual representation; there are two particular themes that I pursue with special vigor, and that define what's important about my story. First, the late nineteenth century is the period in which scientists shift from drawing to photographing many natural phenomena, and entrust to machines the observing and recording of visual appearances. This story is not a simple one: the virtues of cameras and the vices of eyes were not as clear during the shift as they seem a century later. By reconstructing the details of this shift in eclipse observation, I hope to reveal how scientists weighed the merits of different types of observing techniques, and how the gulf between human and mechanical observation developed. Second, I argue for the importance of attending to the material life of pictures. Contrary to the tendency to treat "images" as the visual equivalent of words or ideas, separable from the media in which they appear, this book argues that one must travel through the printer's studio, the lithographer's stone, and the printing press to understand how images take their shape. Following this story to its end requires that we leave the field, and follow pictures as they are studied, assembled into composites, and printed. It also requires examining the interplay between printing technologies and observing technologies, for developments in the printer's shop could affect the way astronomers thought about and used instruments in the field.

SCIENCE AND IMPERIALISM

The relationship between science and European imperialism has also been the subject of rich studies in the last two decades. The late nineteenth century saw the development of the modern scientific world, divided into disciplines, conducted in universities, laboratories, and observatories, supported by governments and philanthropies, organized and publicized by scientific societies and professional associations. It also saw the dramatic expansion of European empires, and the incorporation of most of Africa, India, and Asia into colonies managed from London, Paris, Berlin, Brussels, or Washington. The literature on science and imperialism asks how these two phenomena were related, how scientists put knowledge at the service of empire and exploited research opportunities created by imperial policies and ambitions. Daniel Headrick's *The Tools of Empire* argued convincingly that science and technology were essential resources for European imperialists, and made possible expansion into Africa and Asia.[15] Following Headrick's lead, a variety of scholars have explored the pursuit of science in the colonial context. Some of this work was patterned after studies of the politics of British and European science, and examined the ways scientists pursued careers in the colonies, developed research opportunities abroad, articulated service roles for science in the colonial context, and built institutions.[16] Others studied disciplines, such as tropical medicine and ecology, that were not transplants from the metropole but evolved in response to local needs.[17] Edward Said's *Orientalism* and subsequent work in cultural studies and subaltern studies inspired others (and Others) to probe the ways in which fields such as anthropology defined colonial peoples and territories.[18] Finally, a few scholars have turned the tables and examined non-Westerners' responses to European science.[19]

Victorian eclipse expeditions traveled through worlds shaped and ordered by British imperialism, the expansion of European technological systems and economies, and the diffusion of Western institutions. Eclipse expeditions thus provide excellent opportunities for exploring the ways in which scientists forged connections between astronomical and imperial interests, and used colonial resources for scientific purposes, and for studying the two issues in depth. First, they let us see how scientific and colonial culture intersected, and the ways in which the material and social lives of British colonial administrators, military officers, and others affected life in the field. The documentary record also provides us with vivid comparisons of British and indigenous reactions to totality, which highlight the perceived (or assumed) differences between European and non-European peoples. Second, the conduct of eclipse field-work provides us with an opportunity to explore the deep connections between the expansion of railroad and telegraph lines, surveys and military jurisdictions, colonial economic and political institutions, and the practice of science in the field. Indeed, the infrastructure of imperialism made it possible for astrophysical fieldwork to be conducted. The connections between science and imperialism are to be found not only in the alignment of political programs, the structure of careers, or the texts and ideas that define the European self and colonial Other: they can also be traced into instruments, and mapped into the field.

Organization

The book is organized as follows. Chapter 2 is devoted to an analysis of eclipse expedition planning and organization, and the broader social and cultural context in which that work took place. So many decisions about observing methods and instruments were made in the months before parties set sail that we cannot understand eclipse fieldwork in isolation from eclipse planning. Further, the large-scale social forces that transformed Victorian science also shaped expedition planning. The professionalization and specialization of the sciences, the emergence of new disciplines, the growth of service roles for science, and the growing importance of government funding for the sciences were all reflected in the composition of planning bodies, the way they operate, and the expeditions they assemble.

Chapter 3 follows expeditions into the field. It analyzes writings about eclipse expeditions, and the literary conventions and economic pressures that shaped them, but its main purpose is to reconstruct the experience of fieldwork. This method is gently patterned on Bruno Latour's notion of "following the actors around," to see how they work and live. Here we follow eclipse parties as they travel to their destinations and prepare themselves for the great event. The work of getting to field sites, setting up camp, preparing or repairing instruments, and training volunteers and assistants was extremely hard work; fortunately, it was all described in great detail in notebooks and publications. No study of eclipse expeditions would be complete without a description of what it was like to stand in the shadow of one of Nature's most dramatic events, and to make observations of an eclipse—or be prevented from seeing it by clouds or rain.

Chapter 4 examines the history of visual and photographic observations of the solar corona. The invention and diffusion of photography is one of the most important events in the history of the nineteenth century; certainly this common reference point of technology and culture is one of the things that make the Victorian era seem familiar to us. Eclipse observation taxed both astronomers and cameras to their limits, and raised questions about the trustworthiness of imaging technologies that didn't exist in many fields. It is also notable because problems of reproducing photographs and drawings for publication turn out to be almost as important as producing them in the field. The relationship between original and reproduced images is one that has received little attention in the history of science, but I hope to make the reader see its importance.

Chapter 5 explores the deep links between astrophysical fieldwork and the material culture of British imperialism. The British empire was an exceptional bricolage with countless local mutations spread over a significant portion of the globe, connected by an infrastructure of steamships, railroads, canals, telegraph lines, repeating rifles, and sanitation systems. If Waterloo was won on the playing fields of Eton, one might say that India, South Africa, China, Kenya, and Ceylon were won in the machine shops of Lambeth, the arsenals of Woolwich, and the physics laboratories of Cambridge and Aberdeen. Technological systems shaped and bound the empire together in ways that historians are only beginning to catalog and understand. The technological character of the late Victorian empire also allowed scientists to turn imperial possessions and protectorates into spaces for and spaces of scientific inquiry. This is nowhere better revealed than in eclipse expeditions. The material culture of empire determined the ability of astronomers to go into the field and study the sun—to turn the empire into their observatory.

Planning Eclipse Expeditions

Introduction

The Royal Astronomical Society's monthly meetings were held on the second Thursday of every month. Before the meetings, the Club of the RAS, an informal group of officers and other active members, dined together at Freemason's Tavern. On the evening of 11 November 1870, the talk around the table focused on the solar eclipse of 22 December, which would be visible in Spain, Gibraltar, Algeria, and Sicily. Some of the diners had observed an eclipse in Spain ten years earlier, and had tried to get the government to support an expedition to observe the December eclipse. Eclipses had always been exotic astronomical phenomena—the beauty of the eclipsed sun was preceded by an hour of strange atmospheric and meteorological phenomena—but their scientific value had not been very high until the 1860s. The rapid growth of astrophysics in the 1860s, advances in photography, and the discoveries made at earlier eclipses had turned eclipses into opportunities to observe the sun's corona and prominences, analyze the chemical composition of its outer atmosphere, and search for intramercurial planets. Earlier in the year, the Royal Society and Royal Astronomical Society had both started planning expeditions, collecting information about possible observing sites and calling for volunteers. They combined their efforts in the summer, but a proposal for a government grant and transport aboard a navy vessel had been denied. In October, however, word reached Astronomer Royal George Airy, who had been a major figure in these efforts, that a ship might be available after all. A second committee rewrote and resubmitted the application, and leaders from the two societies made the case for government support of an expedition to the prime minister. Some members of the club worried that that the government's delays made it impossible to organize a proper expedition. Others saw official reluctance as a bad omen for British science generally, proof that efforts to increase public support for science still had a long way to go. Speculation about the government's plans was running high, and some weighed the possibility of organizing private parties to Spain. During dinner, though, "news was brought by Mr. Lockyer . . . that the Treasury would concede all that is required," Airy wrote. "About 8:30 I announced it to the Society."[1]

The Royal Astronomical Society's meetings took place in rooms the society occupied in Burlington House, a neo-Gothic quadrangle near Piccadilly Circus. The Royal Academy of Art, the Linnean Society, and several other scientific societies were also located in Burlington House; the building also lent its name to Burlington Arcade, an adjacent covered walkway with fashionable shops. The RAS was the center of the British astronomical world: its journals published the leading research, its membership and prizes defined the community's hierarchy, and its leaders were connected to the Royal Greenwich Observatory, Oxford, Cambridge, and other centers of research and teaching. The news of government support was greeted by the society's members with enthusiasm, even though there was precious little time to prepare: Airy was privately skeptical, and worried that a botched expedition would make it harder to get government funding for research. But he kept his doubts to himself, and "earnestly begged for information as to names" of volunteers.[2]

Thus began the planning for the eclipse of 1870. All scientific expeditions are planned, but because eclipses last only a few minutes, eclipse expeditions were scripted with particular care. Months of work went into selecting observers; choosing field sites; establishing itineraries; deciding on instruments, observing plans and means of recording observations; and negotiating the rights to photographic plates and the publication of reports and popular accounts. So important was this work that Irish astronomer Robert Ball wrote that while the few minutes of totality were

a short time in which to commence and complete an elaborate serious of observations and measurements . . . by skillful organization of the work it is now possible for a corps of experienced observers to effect . . . an amount of careful work that would surprise anyone who was not acquainted with the resources of modern scientific methods.[3]

That "skillful organization" is the subject of this chapter. I first trace who organized these expeditions, and the institutions they created to coordinate and legitimate their efforts. From the 1860s to the early 1880s, no single official body dominated expedition planning; later, it was institutionalized and centralized. The astronomers and astrophysicists who did this work lived mainly in London, Oxford, or Cambridge. Spanning two generations, they were an eminent group, connected through a network of committees, overlapping research interests, and shared political views. The first generation came of age in a scientific world dominated by "gentlemanly specialists," dedicated amateurs who pursued their science privately and received limited support from universities and the government. The second was more progressive and activist: they were believers in the value of professional science, the public benefits of generous state support for research, and supporters of scientific methods of analysis and public administration.[4] The second half of the chapter describes what was involved in planning an expedition, when astronomers also had to research potential field sites, choose instruments, and recruit observers.

Expeditions in Context

The eclipse of 1870 was a turning-point in the history of British eclipse planning. The level of state support was unprecedented, delivered on the basis of arguments of national prestige and utility; it set a precedent that would be followed ever after. The joint action by the Royal Society and RAS also set an important precedent that would culminate twenty years later in a permanent collaboration that oversaw British eclipse planning. It was also a transitional moment. The volunteers who came forward were a mix of gentleman amateurs, government-employed scientists, and academics; they were products of—and in some cases, were producers of—a period of tremendous change in the place of science in British society. Likewise the organizers included both astronomy's establishment and its young Turks.

Some of the people at the meeting had been involved in eclipse field-work for decades, and their backgrounds give us a sense of the way in which astronomical careers developed in a period before graduate programs, degrees, and academic positions defined who had prestige and access to resources. Most prominent among them was George Biddell Airy (1801–1882), the elderly leader of British astronomy, a man with powerful connections in the government and military who had observed the eclipses of 1842, 1851, and 1860. The son of a Northumberland farmer, Airy entered Trinity College, Cambridge, in 1819, earned money tutoring his better-off but less talented fellow students, and won the Senior Wranglership in 1823. He married up—his father-in-law was the Duke of Devonshire's private chaplain—and held professorships at

Cambridge until 1835. He then moved to London and became Astronomer Royal and director of the Royal Observatory, Greenwich, a position that gave him great influence within British astronomy for the next half-century. Stern and exacting in manner, he ran Greenwich with an iron hand. As one of the government's advisors on scientific matters, he had the power to secure or deny official support for expeditions, and his directorship of the Greenwich Observatory gave him control over a large inventory of instruments that could prove useful in the field.

EXPEDITION PLANNING 1842–1870

British astronomers observed total solar eclipses in the 1840s and 1850s, but it was not until the 1860s that anything resembling a national effort to organize eclipse expeditions came to pass—one in which planning was done by a central body, observers were recruited from a variety of institutions, and funding or logistical support was provided by the government. State support for science was not very easy to come by in the 1840s and 1850s, and eclipses were not very appealing scientific subjects. They were interesting and exotic, but the payoff for observing an eclipse was not very clear: there were some interesting phenomena associated with eclipses, such as Baily's beads, prominences, and the corona, but their importance was unclear in a period in which celestial mechanics dominated astronomical research. The eclipses of 1842 and 1851 were widely observed because they occurred over Europe; there was little incentive to travel to Arabia in 1843, or to Latin America in 1853 or 1858.[5] Official support was strictly limited: the Royal Society and Royal Astronomical Society provided background information, printing "suggestions to observers" and maps of the path of totality, but they did not endorse specific observers' plans, offer financial help, or try to regulate the work of observers.[6]

Airy set the precedent of state support for eclipse observation in 1860 when he organized an expedition to observe the eclipse in Spain. That expedition was organized by a very small circle consisting of the Astronomer Royal, Warren De la Rue, and William Huggins. Warren De la Rue (1815–1889) had enjoyed a triumph at the eclipse of 1860: he succeeded in taking the first photographs of the eclipsed sun. De la Rue's most important scientific work was with sunspots and solar photography: his stereoscopic photographs proved that sunspots were depressions in the sun's atmosphere, and from the early 1860s he directed the photographic "sunspot patrol" at the British Association–sponsored Kew Observatory in London. Like many astronomers of his generation,

De la Rue was a businessman who pursued astronomy as a hobby. The son of a noted London printer and bookbinder, he had been privately educated and had traveled widely, and he had received his introduction to astronomy from James Nasmyth, like himself an engineer and inventor. His mechanical talents were not confined to the observatory: as a youngster he designed an envelope-making machine and other pieces of equipment that were used in the family business. William Huggins (1824–1910) likewise had a business background. He took over his family's silk and linen business in the 1840s, oversaw its rapid expansion, and made enough money to retire to his Tulse Hill estate and devote himself to science. He became interested in astronomy in the 1850s and turned to astronomical spectroscopy in the early 1860s.

Airy's work as an expedition-builder may seem paradoxical in light of traditional perceptions of him as an opponent of state support of pure science.[7] Airy justified his and the government's involvement by defining limited roles for both in the expedition. The eclipse of 1860 was close enough to be attractive to a large number of astronomers, if only the way were paved for them: commercial transportation to and within Spain was still rare and costly, customs taxes were high, and it was not regarded as a very interesting or attractive destination. Airy therefore requested that the Royal Navy assign a ship to carry astronomers to Spain, and that the government arrange that parties be waived through Spanish customs. Airy also acted as a clearinghouse between the official world and prospective observers, soliciting information from volunteers about their schedules and observing plans, passing on information about sites, weather, and accommodations, and discouraging anyone "who merely wished to see the picturesque" effects of the eclipse.[8] This was not an unusual pattern of action for British leaders of science in this period. Joseph Banks served as an advisor and coordinator of research, placing naturalists on expeditions and scientific investigation in the orders of ships' commanders, rather than—with the exception of the *Investigator*—creating and managing expeditions. Airy's contemporary Roderick Murchison worked in a similar fashion. He issued suggestions to travelers, placed protégés on expeditions and surveys, and acted as a clearinghouse for reports from military officers, diplomats, and others, but he did not build expeditions.[9]

The 1860 expedition was a great scientific success: De la Rue photographed the eclipse and confirmed the solar origins of the prominences. The next eclipse that involved Airy occurred in 1868. The path of totality stretched from German East Africa, through southern India, and down into Indonesia. Airy, De la Rue, and Huggins served as advisors to Major James Francis Tennant, a member of the Royal Engineers in India and a former government astronomer of Madras, who did most of the organizational work for the eclipse. He spent his 1867 leave in England visiting Cambridge professor George Gabriel Stokes, Huggins, and De la Rue, working with them to draw up an observing program consisting of spectroscopic analysis of protuberances, photography, and polariscopic observations. Airy reviewed the plans and wrote to the Indian government in support of the expedition. The India Office agreed to share expenses with the British government, with instruments purchased for the eclipse promised to the Madras Observatory.[10]

GENTLEMEN, SOCIETIES, AND
EARLY VICTORIAN SCIENTIFIC CULTURE

This kind of small-scale, informal organization was characteristic of science in a period dominated by amateurs, and had only slowly developed an infrastructure and service role in an expanding, industrializing society. When Airy graduated from Cambridge, there were no academic or corporate research laboratories, few teaching positions, a handful of scientific journals, and no widely accepted arguments in favor of industrial or government support for scientific research. Instead, science was practiced largely (but not exclusively) by wealthy gentlemen and by itinerant lecturers. By the time the Crystal Palace exhibition opened in 1851, the scientific world had grown larger, more diverse, and more structured. Science teaching and research were conducted in Scottish universities, in upstart schools in London, and in countless private lecture halls. Scientific culture was not strictly hierarchical, with science diffusing from the universities down into society: several overlapping and competing scientific cultures had their producers, constituencies, and even world-views. These developments not only changed science significantly in the first half of the century: they also foreshadowed the specialization, professionalization, and debates about state funding for science, and the place of science in national life, that would so deeply affect science in the fifty years between the Crystal Palace exhibition and the death of Queen Victoria.

Early-nineteenth-century British science lacked the kind of institutions and resources lavished on German and French savants, whose technical schools and research universities attracted students from all over the world (something that British scientists never tired of pointing out). But it was connected to the larger society in many ways. A visit to the British Museum of the 1840s offers a view of the way science intermingled with art and politics in both high and low culture. The scientific exhibits were not kept separate from portraits and sculpture. Upon entering the museum, you first went through a main hall containing statues of dramatist William Shakespeare and botanist Joseph Banks, a stuffed hippo, gilt figures from Burma, a stuffed llama (sent from South America by Charles Darwin himself, according to a guidebook), pieces of Hindu sculpture, and three different species of stuffed rhinoceros. Upstairs you could see displays of "artificial Curiosities from the less civilized parts of the world [i.e. weapons, costumes, tools, musical instruments, etc.]," stuffed animals and birds, and displays of shells, all surrounded by portraits. In adjacent rooms were cases of reptiles, sea animals preserved in alcohol, insects mounted on cards, minerals and ores, and fossils. From there you moved on to the sculpture rooms. The number of annual visitors to the General Collections increased by a factor of twenty between 1805 and 1875, a small indication of the steadily growing interest in the sciences in Victorian culture.[11]

The amateur ideal. The vision of science as an interest to be pursued by generalists participating in a wider common culture was embodied in the person of the gentleman amateur. For the larger part of the nineteenth century, British science was dominated by men who pursued scientific interests in their spare time, and largely at their own expense. Unlike today, being an amateur did not imply being a dilettante or undertrained, and the term "professional" did not carry with it today's overtones of authority and expertise (and a good income). In fact, the situation was reversed. At the beginning of the century, the only professions were the law, medicine, and clergy, and none of the three had much status. Rank-and-file practitioners tended to be from poor backgrounds, while a profession's leaders were wealthy men respected for their gentlemanly values and broad education, not their technical expertise. The amateur, in contrast, was a figure to be honored. He did not seek financial gain from his studies, and could be trusted to combine intellectual interest and disinterestedness in the right proportions.[12] Of course, this was more a social type than a description of real people. But real people were influenced by it, and tried to mold themselves according to it—or, later in the century, destroy it.

Amateurs did some excellent science. In an era in which the market for science was not well developed, making a living as a scientist was no easy thing, and one was almost doomed to posts that sacrificed original work to teaching and routine. For example, even by Victorian standards, astronomers at the Royal Observatory at Greenwich were famous for working under a strict regime that not only controlled their scientific work but also required them to perform additional duties, such as checking the accuracy of chronometers, the high-precision clocks that were used at sea. The wealthy amateur, in contrast, had "the power of taking up any subject he pleases, pursuing it so long as he believes in the possibility of success." Free from the burdens of routine and possessed of "sufficient instrumental means," they could be innovators and risk-takers, moving into promising areas with new tools and abandoning worked-out fields before the professionals moved in. (The entrepreneurial metaphors had a real potency, for early Victorian scientists applied laissez-faire political economy to science. Science, they believed, was an instrument of moral and economic improvement, a tool to rational activity, and a bulwark of religion, but like commerce, it was best pursued in private. Government intervention, except of a very limited sort, could do only harm.) This was an idealized vision of amateur life, but there was some truth to it. Astronomy was dominated by amateurs until at least the 1870s, and even later many of the biggest discoveries and most important technical innovations—the invention of smooth-running electric clock drives, large telescopes, and astrophotography—were made by amateurs.[13]

Scientific societies. So gentlemen amateurs could be serious scientists, avid fieldworkers, and technical innovators. They could also be vigorous organizers and promoters of science, and they created a number of societies to advance scientific research and discussion. For more than a century after its founding in 1660, the Royal Society of London had represented all the sciences. The first of the specialized societies, the Linnean, was founded in 1788. New societies popped up like mushrooms in the 1800s: the Geological Society of London was created in 1807, the Royal Astronomical Society in 1820, the Zoological Society of London in 1826, the Geographical Society in 1830, and the British Association for the Advancement of Science in 1831, to name but a few of the more important.[14] Outside London, up-and-coming industrial elites, shopkeepers, and artisans founded and patronized scientific societies. Science offered a model of social activity that was at once meritocratic and democratic, for it required and rewarded hard work and effort but was open to all and showed no respect for inherited wealth and titles. It could smooth class tensions by improving the material condition of workers, by demonstrating the naturalness of the existing social order, by reinforcing piety, by making students aware of the grandeur and complexity of nature and the genius of its maker.[15] Naturalists' field clubs, ranging from expensively outfitted societies to low-cost groups who held their meetings in taverns and fields, also flourished, as the growth of the railways brought the countryside and seashore within easier reach of urban naturalists.[16]

These were serious groups, but sociable too. Many inaugural meetings of illustrious societies were held in taverns or coffeehouses, and there were many more dinner clubs that did not turn into formal societies.[17] Bernard Becker's account of a night at the Royal Society gives us a view of this mix of science and society. Becker arrived at Burlington House shortly before "the fashionable hour of 8:30." He found scientists "chatting pleasantly in groups, or examining specimens and photographs of curious organisms brought by a distinguished Fellow, who is the happy possessor of a sun-picture of the gigantic octopus recently washed ashore at Newfoundland." The fellows were "thoughtful-looking, grey-haired . . . men to whom hard work and constant study have become necessary stimulants," now "bent on enjoying a little bit of scientific dissipation, and resolved to make a scientific night of it." The meeting itself was full of activity: papers, demonstrations of experiments, and displays of new instruments were interspersed with "lengthened and lively discussion" in which the fellows "attack[ed] the subject vigorously." At the end of the evening, "the learned Fellows abandon[ed] themselves to tea and scientific gossip."[18]

Scientific societies provided the resources, standards, and communications networks that are the infrastructure of a scientific community. They promoted research, published journals that communicated the latest findings, awarded medals and prizes, built up libraries and research collections, and standardized nomenclature, instruments, and methods. The Astronomical Society was founded to encourage the business of astronomy, and the Geological Society sought to promote the Baconian method and create a hierarchy among elite fieldworkers, theorizers, and local investigators. They also helped define the place of science in British culture. The British Association for the Advancement of Science, the most influential of the new societies, worked hard to articulate an ideology of science and a definition of its place in society that could appeal equally to landed interests and manufacturers, Anglicans and Dissenters, metropolitans and provincials.[19]

Popular science. A looser network of institutions developed in the same period to popularize science. Mechanics' Institutes flourished from the 1820s, offering applied subjects like drawing and mechanics, and more theoretical subjects such as botany and astronomy. They appealed to artisans and workers bent on self-improvement and social advancement, and they were popular among wealthier patrons because they spread the gospel of respect for authority and established order. A growing circle of instrument-makers, chemists, mineralogists, photographers, scientific illustrators, printers, and booksellers served researchers and better-off students of science.[20] Itinerant science lecturers offered everything from broad, fifty-lecture surveys of "Natural History" or "Mechanical and Electrical Arts" to shorter, more specialized series of six to twelve courses on subjects such as freshwater fishes, photography, applications of electricity, and civil engineering. (Nineteenth-century London, which was building bridges, sewers, new buildings, and railroads at a frantic pace, was an ideal subject for lectures as well as a market for them.)[21] A growing number of books and magazines explained the mysteries of natural philosophy, the wonders of the steam engine and pneumatic tube, or the latest expedition to Africa. These ranged from heavy, expensive volumes for wealthy audiences to cheap books for the masses. A few entrepreneurs tried their hands at popular science magazines, but the market in the 1830s and 1840s was still too small for such specialization. Instead, articles on the physical sciences, technology, natural history, and exploration appeared in popular magazines like the *Edinburgh Review, Leisure Hour, Good Works,* and *Cornhill Magazine.* Some were published versions of popular lectures, filled out with interesting woodcuts, but others had as much technical detail and intellectual sophistication as articles written for specialists.[22]

Finally, there was a scientific underground of medical schools, Chartist-founded schools, and university student societies in which far more dangerous and radical ideas flourished. In this milieu, biological theories, philosophy, and experimental science were the basis for critiques of the established order.[23] Materialism was used by opponents of Church privilege to argue that there was no God, and thus the Church was a drain on the nation's wealth. The Lamarckian claim that organisms could pass on acquired characteristics to their offspring was invoked in defense of working-class and women's education. Evolutionary theories arguing that cooperation rather than Malthusian competition is the driving force behind progress underwrote a vision of a democratic England run by shopkeepers, clerks, and craftsmen. Nebular cosmology, the theory that the solar system evolved out of a cloud of luminous fluid, was linked to Benthamite lessons about materialism and social progress.[24] For Yorkshire radicals in the late 1800s, Mechanics' Institutes were not only incubators for socialist thought and disseminators of evolutionary theory: as cooperative institutions themselves, they were also exemplars of worker cooperation, proof that their teachings could be put into practice.[25] This kind of science didn't horrify just conservatives: moderate savants worried that evolutionary materialism and other wild ideas would undermine their own more limited efforts at scientific and social reform.

EXPEDITION PLANNING 1870–1883

This was the world that Airy, De la Rue, Huggins, and their fellows made and inhabited. But by 1870 there were younger members of the RAS who were deeply dissatisfied with the state of British science, and were agitating strongly for reform; they became some of the most active eclipse expedition planners, and their reformist ambitions shaped that work. Foremost among the reformers was Joseph Norman Lockyer (1836–1920), a young civil servant with a growing reputation in scientific London. The son of a surgeon and scientific lecturer, he was a civil servant who had become interested in astronomy and physics and was drawn into the literary and scientific circle organized around publisher Alexander Macmillan. Through Macmillan he came to know novelist Thomas Hughes, scientists T. H. Huxley and Herbert Spencer, and Pre-Raphaelite painter Holman Hunt, and to share their politically and socially liberal views. He became interested in solar physics in the mid-1860s, and quickly became one of its leaders. By the time he made his dramatic entrance into the Freemason's Tavern, he was editor of *Nature* (which he founded in 1869), a fellow of the Royal Society, a member of the RAS Council, and had a gold medal from the Paris Academy of Sciences, but he still had not landed an academic position. Well-paying scientific jobs were few and far between in 1870, and while that didn't disturb independently wealthy men like De la Rue and Huggins, Lockyer needed to make his science pay.[26]

Others moved into action with Lockyer. Arthur Ranyard, a young Cambridge graduate and London lawyer, made himself secretary for the eclipse organizers, collecting responses to a call for volunteers made at the 11 November meeting. Instructions for observers were produced and distributed: Pre-Raphaelite artist John Brett delivered instructions for corona drawing on 18 November, and Stokes followed with instructions for polariscopic observations.[27] Others put together the observing parties. Oxford professor Charles Pritchard and Cambridge tutor William Clifford both recruited groups of students from their universities, and William Huggins organized a more intimate party consisting of himself, Mrs. Huggins, and John Tyndall.[28] The Organizing Committee provided information about transportation and accommodations, and astronomers or students with little experience in solar observation sought guidance on observing plans and choice of instruments. Miraculously, preparations were finished by the time the parties left England on 6 and 7 December.[29]

Lockyer's involvement in eclipse planning resumed the following summer. An eclipse was visible in Ceylon, India, and Australia in 1871, and Lockyer drew up an ambitious program for coordinated work among three parties spread out under the path of totality, while Airy suggested that the India Office underwrite an eclipse party as it had in 1868.[30] Things didn't go smoothly: Lockyer published a *Nature* column aimed at the council of the RAS declaring that "it would be a disgrace to the science of the age" if the council did not sanction an expedition, while the Royal Society refused to apply for government funds without the RAS's endorsement.[31] Finally, the British Association for the Advancement of Science stepped in and secured £2,000 from the Treasury for the expedition. By the end of the month, things were well in hand, and Lockyer was able to crow to Dartmouth astronomer Charles Young: "The government is doing the thing in good style. We have lots of money and as much help as we want."[32] For the rest of the decade, and into the early 1880s, Lockyer led expedition planning, working with the Royal Society on an expedition to Siam in 1875 and organizing private parties to the United States in 1878, Egypt in 1882, and the South Pacific in 1883.[33] Lockyer's dominance came in some measure by default, for the period was one of inactivity in officially sponsored eclipse work. The RAS Council took the position that the 1878 and 1883 eclipses were the responsibility of American astronomers: British astronomers who wanted to observe the 1878 eclipse could count on plenty of help in the field from American authorities, and so did not need RAS support. The transits of Venus in 1874 and 1882 also drew attention and resources away from eclipses. Eclipse expeditions in the late 1860s and early 1870s were sold in part as training grounds for the transit of Venus, but by the late 1870s, with the hard work of the transit expeditions just ahead, eclipses were just a distraction. George Biddell Airy particularly believed that the transits deserved government attention more than eclipses.[34] Further, Lockyer and the issue of state funding of astronomical facilities such as South Kensington were the subjects of numerous attacks in the early 1880s, making support for an expedition led by Lockyer and supported by the Treasury harder to obtain.[35]

Institutional instability and community cohesion. Eclipse planning in the 1870s and 1880s was not regularized or routine; instead, it was often done hastily, by short-lived committees with few resources at their disposal who had to negotiate with parent scientific societies, the Treasury, the navy, and other agencies. But underneath the institutional instability one finds a community of people who worked on planning every eclipse. This group invested British expeditions with a measure of continuity, and with a political significance they would not have otherwise had.[36]

As we have seen, planning for the expeditions of 1860 and 1868 was led by George Airy, Warren De la Rue, and William Huggins. The 1870 RAS–Royal Society committee included those three, plus Cambridge professors J. C. Adams and George Gabriel Stokes, Oxford professor Charles Pritchard, and William Lassell, Norman Lockyer, Edward Sabine, and Alexander Strange. The organizers of eclipse expeditions in the 1860s and 1870s are listed in Table 2. Ten men stand out as particularly important. Airy and De la Rue planned five expeditions from 1860 to 1875; Lockyer was involved in six (though in three he worked almost alone). Adams, Stokes, and Lassell sat on three planning committees, and Huggins, Edward Sabine, William Spottiswoode, and Strange each served on two. Several committee members who served on only one committee were veterans of previous expeditions: Francis Galton and James Lindsay (eclipses of 1860 and 1870, respectively) served on the 1871 British Association committee, and 1870 eclipse veteran Thomas Huxley served on the Royal Society committee of 1875. The overlap between committees was substantial: four members of the 1871 British Association committee were still on the Royal Society's 1870 eclipse committee, and half a dozen were on the RAS Council that had earlier quarreled with Lockyer.[37]

These regulars played the largest role in developing observing plans, choosing sites, and nominating party members. Some were more involved than others. Edward Sabine faithfully attended meetings, but there is almost no extant correspondence with him regarding eclipse expeditions.[38] Airy occasionally removed himself from the proceedings, ei-

TABLE 2

Members of Eclipse Expedition Planning Groups, 1860–1883

	Total	1860	1868	1870	1871	1875	1878	1882	1883
John C. Adams	3			x	x	x			
George Airy	4	V	x	x	x				
John Browning	1				V				
T. W. Birr	1				x				
William Clifton	1					V			
Dallmeyer	1				x				
Warren De la Rue	5	B	x	x	x	x			
Edward Frankland	1				x				
Francis Galton	1				V				
G. Griffin	1				x				
John Herschel	1			x					
J. R. Hind	1			x					
William Huggins	2		V	B					
T. H. Huxley	1					V			
Knott	1			x					
William Lassell	3			x	x	x			
James Lindsay	1				V				
Norman Lockyer	6			B	B	x	B	x	x
William Miller	1			x					
Charles Pritchard	1			B					
Edward Sabine	2			x	x				
William Sharpey	1			x					
William Spottiswoode	2					x	x		
George G. Stokes	3			x	x	x			
Strachey	1				x				
Alexander Strange	2			x	x				
William Thompson	1				x				
A. W. Williamson	1				x				

ABBREVIATIONS: x, participated in planning expedition; B, planned and participated in expedition; V, planned expedition and observed earlier eclipse.

ther out of fatigue, concern that the government would grow tired of sponsoring astronomical escapades, or worry over the short deadlines and potential for lost face that a failed expedition would bring.[39] Other indicators show who in the group was most influential. The 1870 instructions were written by Warren De la Rue, instrument-maker John Browning, and the Manchester photographer and inventor Alfred Brothers handling photography; Huggins, Gladstone, and Lockyer, spectroscopy; Airy, Stokes, and Charles Pritchard, polariscopic observations; and Lassell, Strange, and Dallmeyer, general observations.[40] Six of these men—Airy, De la Rue, Huggins, Stokes, Pritchard, and Lockyer—and expedition secretary Arthur Ranyard were also on an Organizing Committee that took charge of preparations after the November

TABLE 3

Eminence Indicators of Eclipse Expedition Planners, 1860–1883

	Expeditions planned	Officerships			Awards	
		RAS	RSL	BAAS/ other	RAS medal	RS medal
John C. Adams	3	x	x		1	
George Airy	4	x	x		2	1
John Browning	1					
T. W. Birr	1			x		
William Clifton	1					
Mr. Dallmeyer	1					
Warren De la Rue	5	x		x	1	
Edward Frankland	1					
Francis Galton	1					
G. Griffin	1					
John Herschel	1	x	x		2	
J. R. Hind	1	x				1
William Huggins	2	x	x		2	
T. H. Huxley	1		x			1
Mr. Knott	1					
William Lassell	3	x				1
James Lindsay	1	x		x		
Norman Lockyer	6	x		x		1
William Miller	1		x	x	1	
Charles Pritchard	1	x	x		1	
Edward Sabine	2		x	x		1
William Sharpey	1		x			
William Spottiswoode	2		x	x		
George G. Stokes	3		x			
Mr. Strachey	1					
Alexander Strange	2	x	x	x		
William Thompson	1		x			
A. W. Williamson	1		x			
TOTAL		11	14	8	7	6

meeting.[41] The 1871 instructions were largely a revised version of the 1870 plans, with additional instructions written by John Brett and Henry Holiday, both veterans of the 1870 eclipse and friends of Lockyer.[42]

This was a distinguished group. Table 3 shows the number of planners who were presidents or officers of the RAS, Royal Society, or British Association, or were winners of medals from the RAS or Royal Society. The inner circle of planners—those who worked on two or more expeditions—included four winners of Royal Society medals for scientific achievement, five winners of RAS gold medals, and three senior Wranglers. Five were elected president of the Royal Society, five were president of the RAS, three held professorships at Cambridge and three at South Kensington. Even the men who served on only one committee were an impressive lot, with several medals, presidencies, council seats, high Wranglerships, and Cambridge chairs to their credit.

ECLIPSES AND LATE VICTORIAN
SCIENTIFIC POLITICS

A number of eclipse planners were part of the movement to improve the place of science in late Victorian society. Advocates of increased state support argued that the weakness of science education, the flimsiness of its infrastructure for science, the lack of career opportunities, and the failure of industry and government to appreciate and use science would hurt Britain's competitiveness as an economic and imperial power. T. H. Huxley's experience shows what good scientists of modest means had to deal with. Born in 1825, he was a veteran of an around-the-world scientific cruise, member of the prestigious Royal Society, and winner of its Gold Medal. But despite his achievements, after his return to London in 1850 he spent four hard years unemployed, rejected for the few professorships that were available, forced to survive on popular writing and lecturing. Twenty years later, things were improving, but slowly enough for Norman Lockyer to claim that there was still "absolutely no career for the student of science. . . . True scientific research is absolutely unencouraged and unpaid."[43] The government had not been completely ignorant of science. A government-funded Geological Survey was created in 1835, a fund for scientific research established in 1850, and the Kew "sunspot patrol" conducted in the 1860s; the Royal Society tightened its requirements for membership in 1848. Significant in their time, to the next generation these achievements seemed paltry. In the late 1860s writers were sounding the alarm over Britain's industrial decline and strategic vulnerability, and arguing that more support for science was the only way to keep Britain prosperous and secure in a rapidly changing world economy. The size of the scientific community had increased, and a larger number were training on the Continent, where they saw government-supported science at work. The publication of Charles Darwin's *Origin of Species* in 1859 had raised a firestorm of controversy (though perhaps not as much as once thought), and had proven to many scientists the need for greater freedom to conduct their work—a requirement that could be fulfilled only by expanding the infrastructure for science, and making science a career.[44]

Societies. Improving the infrastructure for science meant reforming scientific societies, expanding science education and research, and increasing public support for science. Scientific societies were important as symbols of the state of science, for their connections to the government, and because they were one of the main arbiters of status within the scientific community. The workings of the "X Club," a dinner group consisting of Huxley, John Tyndall, Herbert Spencer, and others, shows how valuable societies could be. The club was famous in late-Victorian circles for the scientific reputations of its members, the power of their allies, and their simultaneous domination of several societies. The Royal Society was a special target of their efforts: from 1870 to 1885, at least one X Club member (and sometimes four or five) served on the council or held office. At the same time, they kept in the public eye delivering lectures, writing books and editorials, involving themselves in the British Association and Linnean Society, and serving on government advisory committees.[45]

Universities. University-level science offered the greatest opportunities for doing pure research, training new scientists, and developing ties to industry and government. It also held great symbolic importance: if the sciences were respected in Oxford and Cambridge, then they would be accepted everywhere. The opportunities and working conditions of university scientists were improving, but slowly. In the 1820s and 1830s, the founding of London University, University College, King's College, and the Polytechnic Institution all expanded facilities for science education in London. In the wake of the growth of global telegraphy in the 1860s (a spectacular demonstration of the practical uses of physics), ten new physics labs were founded around the country, and engineering laboratories and professorships were also created from the late 1870s. Science was also improving its position at Oxford and Cambridge: Oxford built a new museum, physics laboratory, and observatory, created an undergraduate program in natural science, and recruited a number of promising scientists to its faculty in the 1860s and early 1870s; James Clerk Maxwell was appointed professor of experimental physics in 1871, and the Cavendish Laboratory opened in 1874. As good as these gains were, they could still be rather modest. Cambridge's F. R. Hopkins, for example, was "regarded as a kind of biochemical saint" for putting up with a low salary and heavy duties, and many labs were started in university cellars and paid for out of pocket. The economic depression that started in 1873 also hurt developing science programs. Traditionalist college dons also blocked science appointments, strangled new programs, and otherwise inhibited the creep of "Germanism" into English universities.[46]

Public support. Earlier scientists had argued the merits of science's case in terms of its benefits to individual intellectual development and rationality. Events in the late 1860s and 1870s, including concerns over the relative decline of British industry vis-à-vis her Great Power rivals, worries about the public image of science (particularly in the wake of the antivivisection movement), and the repeated failure of initiatives to secure government support for science prompted a shift of rhetorical strategy. Scientists now cast their argument in terms of "collectivism, nationalism, military preparedness, patriotism, political elitism, and social imperialism." Science yielded tangible public goods: telegraphy, chemicals, improvements in the railroad and steamship, public health—all could be traced back to the laboratory. The rise of Germany as an industrial and political powerhouse gave a geopolitical edge to arguments for expanded funding for science: it supported science systematically and generously. More generally, scientific habits could teach citizens to act rationally, and allow technical expertise to be applied to national problems of increasing complexity and recalcitrance.[47] (This same mix of concern over national decline and faith in professional expertise also influenced the eugenics movement, founded to reverse the decline of the British race brought on by the effects of industrialization. In its heyday, its supporters were engineers, accountants, scientists, doctors, and lawyers, for whom it served as both "a legitimation of the social position" the professional class held and "an argument for its enhancement." Ultimately, eugenics' supporters envisioned a society that would be managed by experts like themselves, with individual abilities and talents identified [and certified] by examination.)[48]

Amateurs and professionals. The reform and improvement of the sciences also required significant changes in the culture of science. The first was the opening of a rift between science and religion. Earlier generations of reformers had worked hard to maintain a balance of power between the scientific and religious establishments. The X Club generation attacked that entente with relish. While they contended belief and scientific thinking were incompatible—as Huxley put it, one cannot be "both a true son of the Church and a loyal soldier of science"—they were keener to overthrow the clerical domination of schools, museums, and public institutions, and capture those institutions for themselves.[49]

Concentrating power in the hands of reformers and their allies also required taking it away from less qualified amateurs and building a distinction between "professionals" and "amateurs" that recognized the superiority of the first over the second. But the definition of "amateur" and the ways the distinction played out are quite complicated. A number of late-Victorian scientific controversies pitted professionals against aristocratic dilettantes, casual scholars who claimed to stand equally with professionals, and industrialists who were mainly interested in the commercial benefits of science were easy to criticize. But reformers did not oppose the involvement of all "amateurs" in science, and such distinctions seem to have mattered little in the conduct of everyday, "normal" science. Charles Darwin, who was lionized for his fearless devotion to the truth in the face of clerical hostility, lived off his inheritance and investments and never held an academic post; likewise, four members of the powerful reformist X Club were amateurs.[50] In astronomy, talented amateurs served on the Royal Astronomical Society's specialized committees and ruling council, won elections to the presidency, and were awarded the Gold Medal well into the 1920s. But their numbers were declining, and the social composition of the society shows both the broad trend and its gradual character. Between the 1840s and the late 1860s, 21 percent had been military officers, and 15 percent had been clergymen; between the 1870s and the early 1890s, those figures drop to 10 percent and 12 percent, respectively. Amateurs and professionals could collaborate outside the RAS as well: the British Astronomical Association, which was founded in 1890, was created for amateurs but run by an alliance of amateurs and professionals, many of whom were already in the Royal Astronomical Society.[51] Professionalization movements outside the sciences worked to further erode the status of amateur scientists and the opportunity to pursue science as a serious hobby. Late-Victorian reform of the clergy led to a renewed emphasis on pastoral work, and reforms in the army and navy made it harder to place social life and hobbies before training and duty.[52] The fracturing of the scientific world, and the redefinition of its relationship with mainstream society, was paralleled by changes in the world of letters. Just as specialized disciplines recast a scientific landscape formerly divided into supercontinents of natural history, astronomy, and the like, so too did academic reform and the growth of professional values in history, literary criticism, economics, and the social sciences all help turn the "man of letters," writing on a variety of subjects for a general audience, into the modern "intellectual," alienated from middle-class society, withdrawn into the ivory tower, pursuing knowledge rather than economic reward.[53]

Eclipse planners in the 1860s and 1870s are notable for the degree to which they were involved in all these reform efforts. Lockyer, De la Rue, and Strange were vigorous advocates of public funding of astrophysical research in the early 1870s, a subject of bitter controversy within the RAS. Their allies in this fight included fellow insiders Adams, Huggins, Lassell, and Stokes, as well as Charles Pritchard. Their principal opponent, RAS secretary Richard Proctor, was not involved in expedition planning (though two other opponents, John Browning and T. W. Birr, were on the 1870 committee).[54] Huggins, Edward Frankland, Lockyer, Stokes, and Huxley all served on a British Association committee on state funding for science in the late 1860s. Lockyer, Stokes, and Huxley were also members of the Devonshire Commission on scientific education, which advocated increased government support for scientific research and education; Lockyer was its secretary.[55] Finally, this group was loosely connected to the X Club, as three one-time planners, Frankland, Huxley, and Spottiswoode, were members of the club.[56] Some fieldworkers were also reformers: 1870 eclipse veteran Henry Roscoe spent decades arguing the case for government support of science.[57] To this group, eclipse expeditions were not just expressions of international scientific eminence, as they were for laissez-faire opponents of public support of science. They were also examples of successful government support of science, the kind that contributed to and showed off British scientific glory, and which if applied at home to laboratories, observatories, and graduate programs could bring permanent improvements in Britain's imperial stature and industrial power.

Science in the colonies. There was one other venue in which British scientist-reformers linked national concerns to programs for scientific advancement: in Britain's growing overseas possessions. The expansion of the empire opened up whole new regions of the world to scientific research, and created problems that begged for scientific solution. Most of the science that British colonial governments supported had this practical aspect to it: budgets were too small to invest in science that didn't promise some return. For scientists, colonies offered opportunities to develop entirely new lines of research, extend old lines into new territories, build institutions and careers, and demonstrate the utility of science to audiences back home. These efforts also created an infrastructure and resources that were important for the success of eclipse fieldworkers.

Science was part of the wider colonial culture in the same way that it was part of domestic British culture. The large amounts of time spent in the field by officers in colonial surveys, the civil service, and military created opportunities for amateur collecting and ethnography. Professional hunters likewise saw themselves as amateur naturalists and anthropologists, thanks to their extensive contact with native peoples and wild animals.[58] Coast and geological surveys, astronomical observatories, botanical gardens and agricultural research stations, natural history museums, supported by a mix of governments, private citizens, and trade associations, offered opportunities for British scientists to establish their reputations abroad. Migration wasn't necessarily permanent: those savvy enough to leverage their local connections into relationships with metropolitan scientists (by sending them hard-to-get seeds, exotic native artifacts, new fossils, and so forth) could eventually win positions back in Britain. And of course there were scientific expeditions, which ranged from small parties, spending years in the field, to huge groups going into the field for a few days. Colonial expansion and expeditions worked together in two ways. Expeditions could be advance agents of colonial intrusion, providing military planners or colonial settlers with intelligence on the value of unoccupied lands and the character of its native inhabitants. Or expeditions could follow the flag, relying on colonial resources to get into the field, or to have changed it.[59]

The cases of colonial geology, ecology, and tropical medicine illustrate the ways that ambitious scientists cultivated opportunities for research in Britain's colonies. Geology required little up-front investment—a hammer, magnifying glass, a few cheap chemicals—and it offered both practical gains for miners and landowners, and the intangible rewards of fresh air and exposure to nature that made hunting appealing. The Ordinance Survey, created in 1835 to survey and assess the mineral wealth of Great Britain, encouraged colonial geology almost from its founding. Colonial governments liked mineral surveys, as few things could attract settlers, investment, and hard currency faster than the prospect of rich mineral resources. Geologists further argued that surveys could direct settlers into territories rich in resources but as yet unclaimed (by Europeans, anyway), leaving the mineralogical wastelands to Britain's competitors. The modest cost of a survey was trivial compared with the tons of precious ores and long-term geopolitical benefits it could yield. Behind many of these efforts was Sir Roderick Murchison, one of Britain's leading geologists and a tireless advocate of the colonial uses of geology and geography. Murchison lobbied colonial governments to create surveys, which he then staffed with his students, who in turn provided both mineral surveys for colonial authorities and data for Murchison's own theories.[60] Conservation likewise developed in response to practical needs of colonial governments. Studies of the relationship between deforestation and local climate were first pursued on the island colonies of St. Helena and Mauritius, provisioning stations for British and French navies that had been severely strained by plantation agriculture and deforestation. The discipline blended observations of local climate and rainfall with "a whole range of intellectual concepts developed in regions as diverse as France, the Cape Colony, China, and Malabar," at a time when "no comparable intellectual development of an environmental consciousness had yet emerged in metropolitan Europe."[61] Tropical medicine developed in India early in the nineteenth century, when Scottish-trained doctors crossed the latest Western medical ideas with Indian practices to produce a theory of hygiene that stressed environmental causes of disease and the need for Europeans to adopt local diet and clothing. Later physicians argued that native practices were unhealthy and unscientific, and advocated segregation of Europeans to protect them from the unhealthful habits and endemic diseases of indigenous populations. The evolution of tropical hygiene into tropical medicine in the 1880s and 1890s—with the schools of tropical medicine, journals, societies, and the application of statistics and laboratory studies—was the last step in the intellectual segregation of Western and Indian medical thought and practice.[62]

The impact of colonial science and Western technology was mixed. On the negative side, colonial science was almost always defined by its advocates as a tool for rationalizing and improving colonial political and economic activities, rather than indigenous industries and practices. Botanical gardens focused on plantation crops rather than indigenous foods, and agricultural societies were usually dominated by plantation owners.[63] Anthropology and tropical medicine likewise were not about dialogue, or communicating native ideals and perceptions: they were about speaking about and for colonial subjects. The impact of Western technology on imperial possessions is just as complicated. Military and transportation technologies were critical for multiplying and projecting European military power in the nineteenth century. On the other hand, the transfer of technology under colonial governments was just the latest in an often-substantial traffic in Western technologies into Asia and Africa from the sixteenth century onward, and European systems were sometimes used by indigenes for their own purposes. In India, railroad lines were intended to assist in the exportation of raw goods, and to project military strength; but they also resulted in a dramatic increase in pilgrimages to holy sites and contributed to the coalescing of an Indian national identity and the independence movement.[64] Native peoples were part of the colonial scientific working class, serving as guides, trackers, artists, computers, and technicians. The Great Trigonometric Survey had a sizable staff of Indian assistants, and the Royal Engineers trained Indians in covert surveying methods for use in areas forbidden to Europeans. Likewise, British botanists relied on skilled Indian artists to provide illustrations for plants. This work was crucial to the success of European scientific efforts, and it provided a limited, practical training to native scientists.[65] In India, native elites also created a scientific world that mirrored the one built and populated by Europeans. European doctors, mercenaries, engineers, and others had served as advisors and courtiers in Indian princely courts since the 1500s.[66] In the colonial period, native patrons endowed astronomical observatories and hospitals, and pushed for Indian access to Western scientific education.[67]

EXPEDITION PLANNING 1885–1905

A new community of eclipse planners began to form in 1885, when the Royal Society resurrected its Eclipse Committee to advise the New Zealand government on observing the eclipse of 1885 and to plan an expedition to the Caribbean in 1886. In 1888 the Royal Astronomical Society created the Permanent Eclipse Committee to develop "more perfect arrangements" with a more generous planning window, and to secure "the advantage of continuity" in plans from one eclipse to another.[68] The PEC's first task was to make plans for the eclipse of 1889, visible in South America and western Africa.[69] A larger group, formed at the suggestion of Government Grants Committee chairman William Clifford (a former Cambridge tutor and organizer of one of the eclipse parties in 1870) and consisting of the PEC, Royal Society eclipse committee, and Solar Physics Committee (the group that oversaw the South Kensington Solar Physics Observatory), planned the expedition of 1893 to west Africa.[70] That structure proved too large to operate effectively, and its temporary character also now seemed inefficient; as a result, the RAS and Royal Society agreed to form a smaller, permanent planning committee in 1894. As it turned out, half the members of the PEC already consisted of fellows of the Royal Society, so it was simply renamed the Joint Permanent Eclipse Committee.[71]

TABLE 4

Members of Eclipse Expedition Planning Groups, 1885–1886

	Regular session	Subcommittee sessions	Total
Norman Lockyer	9	7	16
William Christie	10	4	14
George G. Stokes	13	0	13
William Abney	8	4	12
Arthur Schuster	6	4	10
Warren De la Rue	8	0	8
Leonard Darwin	3	4	7
Michael Foster (Sec. RSL)	5	0	5
H. H. Turner	1	4	5
Lord Raleigh	4	0	4
John Evans (Treas. RSL)	1	2	3
E. Walter Maunder	1	2	3
T. E. Thorpe	1	2	3
Stephen Perry	1	1	2
William Huggins	1	0	1

The advantage of institutional continuity. The JPEC was invested with three powers that no earlier planning body possessed. It was given control over the publication of expedition reports (something that caused confusion in the past). It could apply directly to the Government Grant Committee of the Royal Society for funds.[72] These powers had the effect of making the JPEC the nation's official spokesman on eclipse matters and gave it a monopoly over government funds and favors.[73] Most important, it was given the power to purchase and manage instruments, which had previously been "distributed . . . to various irresponsible bodies or persons."[74] This provision had the subtlest and most far-reaching consequences of all. It allowed the committee to develop a stock of high-powered, specialized instruments unique in Britain, and in concert with its other new powers gave it the ability to mandate their use over several expeditions or even decades.

Community composition. As the transition from the PEC to JPEC suggests, there was considerable overlap between membership in these three different groups. An analysis of attendance at meetings can tell us who the group's core members were. (See Table 4.) Planning for the eclipses of 1885 and 1886 was dominated by a handful of men. William Abney, William Christie, Norman Lockyer, G. G. Stokes and Arthur Schuster were the most regular attendees. (Lockyer and Stokes, and less involved members Warren De la Rue and William Huggins, were veterans of eclipse planning efforts in the 1870s.) The regular meetings focused on financial matters, diplomatic arrangements, logistics, and recruitment of expedition members. At the subcommittee meetings, where the serious technical decisions about instruments and observing programs were made, Leonard Darwin, Edward Maunder, Stephen Perry, T. E. Thorpe, and Herbert Hall Turner emerged as active players. These men, and Abney, Christie, Lockyer, and Schuster, were the central figures in the Royal Society's planning effort. Similar records for the JPEC show that this same small group—with the exceptions of De la Rue, Stokes, and Perry, who died in 1889—dominated eclipse planning in the 1890s. As before, Huggins was a member of the JPEC in name only, coming to one meeting in twenty years. Christie, Common, Lockyer, Maunder, and Turner continue to be important players, and they are joined by Dyson, Fowler, Hills, Maw, MacMahon, Newall, and Johnston Stoney. Others withdraw somewhat from eclipses. Leonard Darwin and Arthur Schuster are active in the Royal Society committee, but not the JPEC, and Abney slows down somewhat as well. High attendance correlates strongly with participation in expeditions: membership in the JPEC became a ticket into the field.[75] (See Tables 5 and 6.)

All the principal figures lived within an hour or two of London. The Astronomers Royal for Ireland and Scotland, while they were members of the committee and corresponded with its secretary, were too far away to attend its meetings. They had other points of overlap, serving on other

TABLE 5

Principal JPEC Figures: Years Joined and
Meeting Attendance, 1894–1909

Name	Year Joined	Attendance Rate
William Abney	1894	28%
William Christie	1894	92%
A. A. Common	1894	61%
A. M. Downing	1894	19%
Arthur Fowler	1894	78%
E. H. Hills	1894	83%
Norman Lockyer	1894	47%
E. Walter Maunder	1894	64%
P. MacMahon	1897	48%
H. F. Newall	1896	72%
Arthur Schuster	1894	14%
Walter Sigreaves	1894	14%
E. J. Stone	1894	54%
G. Johnston Stoney	1894	64%
H. H. Turner	1894	78%

TABLE 6

RSL/RAS/JPEC Expedition Membership, 1886–1900

Name	1886	1887	1889	1893	1896	1898	1900
William Abney							
William Christie					x	x	x
A. A. Common					x		
Ralph Copeland		x				x	x
Leonard Darwin	x						
A. M. Downing							
Frank Dyson							x
John Evershed							x
Arthur Fowler				x		x	x
E. H. Hills					x	x	
H. Lawrence	x						
Norman Lockyer	x				x	x	x
E. Walter Maunder	x						
P. MacMahon							
H. F. Newall						x	x
Stephen Perry	x	x	x				
Arthur Schuster							
Walter Sigreaves							
E. J. Stone							
G. Johnston Stoney							
Arthur Taylor			x	x			
T. E. Thorpe	x			x			
H. H. Turner	x				x	x	x

RAS committees (the Photographic Committee was popular), meeting at Royal Society soirées, dining together in the RAS Club, even playing golf together regularly.[76] Christie felt that the full committee was "too unwieldy" for most purposes, and he and Edmond Hills met alone to talk over plans for the eclipse of 1898.[77] Finally, despite the year-long memberships, half of the JPEC's founding membership was still on the committee in 1902.

Like the previous generation of planners, the group that dominated the JPEC was quite distinguished, but the education and careers of these principal figures shows that the standards for admission into elite scientific circles had changed in twenty years. Previously, there was a healthy representation of Cambridge alumni, professors, and Greenwich astronomers, but now they dominate the inner circle. George Gabriel Stokes (1819–1903) came from a family of Anglican ministers, was Senior Wrangler in 1841, a fellow at Pembroke College, Cambridge, and Lucasian professor from 1849 to 1903. William Christie (1845–1922) graduated from Trinity College, Cambridge, in 1868, served as chief assistant at Greenwich for a decade, and succeeded Airy as Astronomer Royal in 1881. Edward J. Stone (1831–1897) was Fifth Wrangler in 1859, went immediately to Greenwich, spent another eight years at the Cape Observatory in South Africa, and from 1878 was director of the Radcliffe Observatory at Oxford. The career of Herbert Hall Turner (1861–1930) paralleled Stone's. After graduating as Second Wrangler at Cambridge in 1882, he spent nine years as chief assistant at Greenwich, then in 1893 succeeded Charles Pritchard (who observed the eclipse of 1870) as Savilian professor of Astronomy at Oxford. Arthur Schuster (1851–1934) was a German émigré who returned to Heidelberg to earn a doctorate in physics, and served as a research assistant in the Cavendish Laboratory at Cambridge before becoming professor of physics and mathematics at Owens College, Manchester, in 1881. Edward Maunder (1851–1928) left Cambridge without taking a degree but spent several decades as an astronomer at Greenwich. In short, eclipse planning became the province of a circle working in Oxford, Cambridge, Greenwich, and South Kensington.

As with the previous generation, the JPEC's members also included some of the most politically powerful men in British science. The JPEC included half a dozen RAS presidents, four secretaries, two treasurers, and four Gold Medal winners. Five of the JPEC's members—Abney, Huggins, Lockyer, McClean, and Schuster—simultaneously served on Board B of the Royal Society's Government Grants Committee, which evaluated research proposals in experimental physics, astronomy, and meteorology. Two other Board B members, William Clifford and Charles Joly, were not on the JPEC but were expedition veterans, and another, Richard Strachey, also served with Lockyer and Abney on the Solar Physics Committee.[78]

The criteria for being part of the astronomical elite changed in a significant way between the 1860s and 1870s and the 1890s: being a professional scientist meant a lot more. Earlier, gentlemen amateurs had balanced the academics in eclipse planning bodies. Now, however, only two important JPEC members did not have connections to academia or the government. One was A. A. Common (1841–1903), who designed some of the best reflecting telescopes of the late nineteenth century. The other was Edmond Hills (1864–1922), an army officer and sometime instructor at Woolwich, who served as the JPEC's energetic secretary. There were a few other gentlemen amateurs or career military men who served on the JPEC, but they were peripheral figures, serving short periods and rarely appearing at meetings. This reflects a larger shift in the balance of power in astronomy: disciplinary control was moving clearly into the hands of professional astronomers. A "brilliant group of semi-amateurs" had dominated the field into the 1870s, producing its most important work and winning its awards. In the last quarter of the century, however, the number of important amateurs declined, and organizational lines more strongly differentiate amateurs from professionals.[79] A few exceptional amateurs managed to remain competitive in an increasingly professional world, and an even smaller number managed to turn professional but were driven to the periphery of the field.[80]

Priorities in Expedition Planning

There were three broad kinds of planning work that had to be done to prepare an expedition. First was the political work of drafting proposals for funding and convincing the government to support an expedition. Second was the logistical work of researching field sites, making travel plans, finding local volunteers, and getting accommodations. Third was the technical work of choosing instruments, deciding what kinds of observations should be made, and who should go into the field. This required information from many different sources, and while we will look at each in turn, in reality work on all three fronts tended to proceed simultaneously. At any given time, an organizer's desk would be piled high with maps, meteorological reports, guidebooks, recent issues of the *Philosophical Transactions*, steamship timetables, and letters from the Foreign Office, Treasury, colonial administrators, and foreign colleagues. Coordinating all these materials and plans was quite time consuming. Norman Lockyer advised that "a month's interval is not too long for these preparations; a fortnight should be considered as an irreducible minimum."[81] In fact, he usually spent several months planning his expeditions. Not all of this work qualifies as "scientific" in the traditional sense, but astronomers themselves realized that settling political and logistical issues was just as essential for the success of an expedition as choosing instruments.[82]

INTERNATIONAL COMPETITION

Why did the government support eclipse expeditions? The rhetoric that astronomers used when seeking funding suggests the government could be moved in part by appeals to nationalist pride and national obligation. Of course, this was a period in which most astronomical research was self-consciously (even aggressively) international in character, and eclipses, the subject of "great attention in all civilized nations," as Airy put it, might at first seem natural candidates for international joint ventures.[83] However, a combination of factors worked to turn expeditions to study these "astronomically and cosmically significant" events (quoting Airy again) into nationalist symbols.[84] Eclipses drew lots of attention from the international press, and an expedition was a highly visible demonstration, both at home and abroad, of scientific maturity and vigor. During planning for the 1870 eclipse, an editorial in the *Daily Mail* probably written by Norman Lockyer warned that "our American fellow-workers in science are here amongst us . . . prepared to learn from our longer experience; but also sharp-eyed to detect the signs of effeteness in the science of the mother country."[85] In an era in which worries about the decline of British science were widespread, Lockyer suggested, failures to support highly public ventures like eclipse expeditions would be costly.

The nature of Continental and American competition varied. French and German expeditions were organized by a mix of scientific societies, government agencies, and leading observatories not unlike that of the British. American eclipse expeditions, in contrast, were local affairs, planned, funded, and outfitted by individual observatories and colleges, or by astronomers borrowing instruments and traveling on small grants from various institutions or agencies. No national-level group had the power, either through control over funds and official favors, or through scientific respectability, to assert a European-style dominance over astronomy and set national standards in practices, personnel, or observing methods. The National Academy of Sciences lacked the power or prestige of the Royal Society of London, nor was the American Association for the Advancement of Science as influential as its British opposite. There was no American equivalent of the Royal Astronomical Society: local groups such as the Chicago Astronomical Society expired within a few years of their founding, and the Astronomical Society of the Pacific consisted mainly of amateurs and was dominated by the Lick. It was not until 1897 that a national astronomical society was founded, and its membership remained anemic for years. The U.S. Naval Observatory provided some instruments and assistance for small parties, but Simon Newcomb didn't wield the Astronomer Royal's power. The wide distances between observatories further reinforced the independence of observatories and astronomers from the directives and controls of national societies and peers. As a consequence, the "local environments" of university and college observatories, the individual initiative of observatory staff, and local funds determined who led American eclipse fieldwork. Some small observatories like Princeton and Amherst organized more expeditions than the Harvard College Observatory, Smithsonian Astrophysical Observatory, or Mount Wilson.[86]

Scientific preeminence was also one of the justifications that nineteenth-century apologists gave for European imperial expansion, and expeditions within one's own territory were vivid demonstrations of a nation's superiority over its colonial subjects. Imperialist rhetoric also shaded into imperial obligation: to have the eyes of the world turned to British territory in the Caribbean or India and not send out an official party in the field was unthinkable. This logic also demanded that even if better conditions were available elsewhere, British astronomers had to work on British territory whenever possible.[87] But there were limits to geopolitical responsibilities: the British government was not obliged to send parties to America or into the colonial territory of other Great Powers.

RESEARCHING SITES

Convincing the government to fund an expedition was child's play compared with the work involved in figuring out where to go, how to get there, and how to survive in the field. The first step was to find out the path of totality. By the late nineteenth century, the mathematics required to divine the location of the "shadow path" was well developed, and predictions to within a few hundred yards were possible, an excellent level of accuracy since the shadow path itself was more than a hundred miles wide. Preliminary computations of the path might be made by an astronomer, working alone with pencil and paper and simplified equations. However, no one made serious plans without the computations, tables, and maps—often combined in a single booklet or large map—published by their nation's Nautical Almanac office. For example, the British Nautical Almanac map for the eclipse of 1868, produced by Superintendent (and former Greenwich assistant) John R. Hind shows the path of totality from Aden to the Torres Straits. The map lists the latitude and longitude of the central path, and the times at which totality would begin and end at various points. An insert shows the path over India in greater detail. Not only are latitude, longitude, and time given, but the location of several towns within the path are also marked. This was the map consulted by George Biddell Airy when weighing observing plans.[88]

A more detailed map of the same eclipse was made by Major James Tennant. He taped together several large maps of southern India, published by the London-based Society for the Diffusion of Useful Knowledge, drew in the path of totality, then circled the towns under the shadow path accessible by rail.[89] (Society maps were popular among eclipse planners: another was used, perhaps by a Greenwich assistant, when researching sites for the eclipse of 1871, also visible in southern India.)[90] The search for sites was the second stage of planning, and often one of the most time consuming. Tennant's map and notes show that an ideal location for an eclipse camp was a small town, easily accessible by railroad, with clear observing conditions and no clouds or dust.

While a few parties would gather together in a single town, there was also an attempt to spread out as far as possible along the shadow path, to minimize the effects of bad weather in any single region and maximize opportunities for detecting changes in the corona. Arguments in favor of this policy were advanced by Warren De la Rue during an (unsuccessful) pitch to Airy for expeditions to North America and Africa to observe the eclipse of 1861. "It is extremely interesting to have observations so distant as those of Vancouver Island and Algeria," he wrote. "It may not occur again for a very long period that a total solar eclipse is so thoroughly observed as that of 1860."[91] In 1871, the RAS–Royal Society joint committee considered asking the Treasury for funds to send parties to India, Ceylon, and Australia.[92] For the eclipses of 1889 and 1893, the RAS and JPEC sent parties to different countries and even different continents.[93] Astronomers also chose stations at different altitudes, to avoid lowland dust or increase the chances that bad weather at one station would not affect another.[94]

LOCAL INFORMANTS

Once it was determined which towns were most easily accessible, astronomers gathered information about weather, transportation, and accommodations at each site. This material came from a variety of sources and was often difficult to collect, since the most valuable intelligence came from European expatriates drawing on personal experience. General information about the weather was collected and published by national meteorological services, but advice about weather at specific sites came from those familiar with the rhythms and quirks of local climates.[95] Colonial army officers, administrators, and engineers were also tapped for information about transportation, road quality, hotels and campsites, information that drew on their own distinctive kinds of fieldwork.[96] The need for local informants and assistants continued when parties were in the field. Men familiar with local customs, the pace of work (James Tennant warned his fellow countrymen of the "inevitable delays of India" as if they were a force of nature), and standard prices for supplies and wages for workers were invaluable to astronomers occupied full-time with building their camps and getting their instruments in shape. They could also help expeditions deal with local bureaucracies and "Custom House officialism," as one astronomer put it. It was one thing to have a promise that tariffs and inspections would be waived on scientific instruments, quite another to actually get one's equipment through customs untouched; likewise, much work could be required to turn promises of discount railroad fares into reality. Locally based Europeans and their native assistants also served as camp managers, hiring servants, overseeing cooks and laborers, keeping accounts, and receiving visitors. They were also best prepared to deal with local superstitions and suspicions, defusing rumors among the populace about the intentions of the astronomers.[97]

Local officials guided visiting astronomers in search of observing stations, repairmen, and other facilities, and they could count on their help being acknowledged, almost ritualistically, in the expedition's official reports. But field relationships were built of more than ritual, and expedition members often expressed affection for volunteers' work and sacrifice. When the H.M.S. *Psyche* ran aground carrying eclipse observers to Sicily, Lockyer and other astronomers sent a petition, modestly titled a "Memorial of the Undersigned Fellows of the Royal Society and Royal Astronomical Society, and Students of Science" but nonetheless apparently quite heartfelt, to the Admiralty asking them to go easy on the captain, who as a matter of course was being court-martialed.[98]

Colonial officials and expatriate astronomers were critical resources, but dealing with them was sometimes a delicate thing: their local connections and access could make the difference between success and failure, but cosmopolitan astronomers worried how and whether they could be handled, and what obligations might be incurred in the process. George Airy's and Walter De la Rue's efforts to get Norman Pogson to make observations in 1871 captures these tensions. As director of the Madras Observatory, Pogson was an obvious choice, and his official position in the Indian government could be useful. Airy thought little of Pogson—he "does not admire photography," Airy complained to John Herschel at one point—but had no choice but to approach him about leading an expedition.[99] Pogson tried to turn their correspondence into a formal collaboration between himself, the Royal Society, and RAS, while Airy wanted Pogson to follow orders but not as an official delegate of the society.[100]

STAFFING THE EXPEDITION

While sites were being researched and government and colonial favors sought, observers were being recruited to go on the expedition. It is here that the effects of institutional developments make themselves felt most strongly. Parties in the 1860s and 1870s had markedly different compositions than those of the last two decades of the century, and the motives for going on an expedition were slightly different, or at least more complex.

Expeditions in the 1860s and 1870s were a mix of amateur and professional astronomers. Whether jockeying for position in the strictly hierarchical but informal world of amateurs, or competing for one of the few paying jobs in astronomy, participation in an expedition could mark one apart from the pack. Going on an expedition was a well-recognized way to make a name for yourself in the sciences: Charles Darwin was elected to the Royal Society while still on the *Beagle*, and by the time he returned to London his reputation as a naturalist was well established thanks to his dispatches from the field. It was more unusual in astronomy, since fieldwork opportunities were more limited, but it could work. The young lord and future earl of Crawford, James Lindsay, was made a member of the RAS after organizing an expedition to the 1870 eclipse, and he was elected president not long after funding a transit of Venus expedition in 1874. Arthur Ranyard made his reputation with the 1870 eclipse, serving first as Norman Lockyer's secretary while the expeditions were organized, then taking over from Airy the job of editing a volume of pictures and accounts of the eclipse. A recommendation letter for Arthur Schuster pointed to "his successful directing of the Siamese eclipse expedition" of 1875 as evidence of his "powers of organization."[101]

In this earlier period, Airy made sure that only responsible and scientifically competent observers were admitted to an expedition. A volunteer who wanted to "observe the general effect" of the 1860 eclipse "on nature itself" was told that anyone "who merely wished to see the picturesque" was not welcome, and others were told that they would be accepted only if they developed definite plans.[102] Airy likewise investigated Tennant's qualifications and character before endorsing his plans for the 1868 eclipse. "Now in all things of this kind," he told De la Rue, "I think that the main difficulty is about the man."[103] He looked at the 1870 volunteers with some alarm. "There are many names I do not know, and some known to me . . . who are hardly likely to undertake observations in the form which we must require," he told William Lassell. "It will be difficult for responsible men in office"—such as Airy himself—"to urge any claims on Government, except for observations of the class which modern science imperatively demands, arranged in the most distinct organization." Airy went on to define what that class of observations meant: "We cannot now recognize any claims for mere general eye-views of the eclipse, or remarks on the picturesque, or even the rise and fall of the red prominences. We must have severe assurances that instrumental observations will be made, under distinct organization, of the points on which great questions are now raised—spectroscopic and polariscopic, and such things as photography can show."[104] While Airy was strict about allowing only scientifically competent and serious observers, he admitted architects, chemists, solicitors, military men, and reverends, as well as practicing astronomers. Occupation was not yet a clear indicator of scientific worth, which is why Airy felt such distress at not personally knowing many of the volunteers for the 1870 eclipse.

This changes completely in the 1880s and 1890s. The monitoring of expedition membership became virtually a nonissue, as both the size of parties and the pool of astronomers considered qualified to go on an expedition narrowed. The Joint Permanent Eclipse Committee banned amateurs not associated with the JPEC from joining expeditions, on the argument that official support could not be exchanged for regulated observation, and that amateurs might try to carry out a "scheme of work it [the organizing committee] has had no opportunity of criticizing." Just in case some were willing to pay their own way, the committee secretary added, "The fact that no additional expense is thereby entailed does *not* affect the question."[105] Likewise, astronomers were also forbidden from taking observatory mechanics and assistants, on the grounds that they were too expensive and had no special skills that could not be found locally.[106]

FINDING GOOD HELP

Eternally burdened with more instruments and observations than they could manage, and faced with the multiple jobs of building and maintaining a camp, setting up instruments, and doing countless odd jobs and repairs before an eclipse, expeditionary forces were always in need of extra skilled labor. Even those just observing the eclipse with a telescope needed an assistant to point and guide the instrument. Eclipse photography required at least two or three people, for it "overtasks one photographer, unless that one be unusually quick in working and calm under great provocation and excitement."[107] Instruments were best handled by people with some mechanical skill, and preferably some familiarity with exacting tasks that could be learned by drill and rehearsal. Amateur astronomers were an obvious potential source of skilled labor, but they were denied positions on official expeditions from the 1880s. This meant that expeditions had to rely on European-trained colonial engineers, railroad workers, military officers, and soldiers and sailors. Military and naval officers were especially valued for their discipline, acquaintance with optical instruments and drill-work, and ability to perform under pressure.[108] As instruments became larger and more complicated, they required more and more assistants. Norman Lockyer, who had led tiny parties in the 1870s, demanded an entire ship be put at his disposal after 1896: "So much assistance is required for the proper manipulation of these [eclipse] instruments," he argued, "that unless a strong staff from a Man of War is available I am afraid that we shall not be able to undertake the work at all."[109] What had been expedient in 1870 was necessity in 1900.

TRAVEL

Travel to and from the site was another major concern of planners. Correspondence about it tended to be of a more limited and straightforward sort, consisting mainly of requests for reduced fares aboard commercial vehicles.[110] Steamship fares could easily have consumed a large part of an expedition's budget, had discounts not been granted. In 1870 the Royal Navy provided the government expedition with transportation worth more than £2,000, more than the expedition's entire budget.[111] The U.S. Navy saved the Lick Observatory $5,000 in shipping and passenger costs on a single expedition.[112] American railroads also provided reduced fares and passes.[113] Maps of rail and steamship lines, and schedules of departures from New York and Southampton, included in many letters outlining plans and options, are often found among the personal papers of eclipse observers.[114] Whether they traveled on commercial or naval vessels, care was paid to providing separate accommodations for gentleman astronomers and paid assistants. Social distinctions that might be less important in the modern observatory still separated gentleman from employee on the high seas. In 1870, for example, Oxford and Cambridge volunteer assistants were placed with the official astronomers, but paid assistants were placed in separate quarters. "I *presume* the [paid] assistants would scarcely be placed with the rest of the party," one gentleman asked Astronomer Royal George Biddell Airy before they left.[115] The

Admiralty, "on account of accommodations," made sure that the entire party would be "on one social footing, and that it is not in contemplation to take any others of a different Class."[116] Similarly, when Edinburgh astronomer Ralph Copeland went into the field in 1900 with an observatory mechanic, the expedition organizer queried, "I presume that he will mess with the First Class petty officers?"[117]

Reliable and modern transportation was important to secure because of the complexity and fragility of instruments. Not only was equipment delicate, but there was lots of it. Warren De la Rue needed two tons of instruments and supplies to make four photographs in 1860.[118] Lick Observatory expeditions traveled with a minimum of eight instrument boxes, each weighing over one hundred pounds.[119] Steamship lines were told to take special care of instrument boxes, and sometimes sealed railroad cars were reserved for expedition equipment.[120] Customs agents were a constant worry, and expedition leaders fretted over their instruments whenever they were out of sight or examined by officials.[121] Porters were another danger. Indian Army officer J. P. Maclear warned Norman Lockyer about "the treatment [luggage and instruments] will receive from the Arabs between Suez and Alexandria," and advised carrying the most fragile items personally.[122] As a consequence, no party went into the field without instruments and observers being heavily insured.[123] The Foreign Office and observatories in countries under the path of totality lobbied for free entry for expeditions. Most countries complied.[124]

PROVISIONS

Finally, parties had to make arrangements for provisions: food, clothing, tents, medicine, and other supplies. Again, European and American informants guided parties. Amherst professor David Todd was told by a missionary to take white clothes, a pith helmet, and a mosquito net to Japan, and "always take an upper room in an inn."[125] Herbert Hall Turner, Savilian professor of Astronomy at Oxford, was advised to take an India-rubber folding bath and four towels to Grenada in 1886.[126] "Bring heavy blankets or rugs, or both, as the nights are extremely cold" in Wyoming, U.S. Naval Observatory director Simon Newcomb wrote to Norman Lockyer. However, he added, "with a flask of brandy in case the water is unwholesome I think health and comfort can be assured."[127] Access to alcohol, of the proper sort and in the necessary amounts, was a matter calling for careful management. "The staple drink of the two senior members of the party is whiskey and soda," Herbert Turner was careful to tell expedition organizers twenty years after Newcomb's as-surance.[128] The amount of food a party carried varied widely, depending on the availability of local supplies, but no expedition was fed badly. As-tronomers were treated to a whirl of dinners by foreign officials and col-leagues, and even in the field expected "all necessaries and a due propor-tion of luxuries."[129] (The only complaints of bad food I have found came from parties in Wyoming, west Africa, and Georgia.) Camping was easi-est in India, thanks to the existence of a well-developed tourist industry and tents and supplies provided by the army and civil service.[130] Food and other provisions could be bought or delivered from Bombay, and plans made against unforeseen dangers. American Elizabeth Campbell, who ran the camps for several Lick Observatory expeditions, had crates of food shipped from Bombay, and asked a doctor for remedies against co-bra bite, tiger attack, sunstroke, and cholera. He recommended scotch.[131]

Conclusion

The Victorians helped create modern science as we know it. Of course science as the rational study of the natural world arguably has its origins in the Hellenic world; scientific societies, journals, and professorships are all innovations that date from the Renaissance; and nineteenth-century German, French, and American scientific communities also pro-duced institutional innovations that influence the sciences to this day. But the nineteenth-century British landscape of institutions and cultural beliefs, the problems scientists confronted, and the role they defined for themselves in society—the places of science, and the place of science in society—are recognizably modern, and still familiar today. Victorian sci-entists were the first to be called "scientists": they coined and popular-ized the term. They oversaw the creation of a dense network of scientific societies, many of which are still world leaders in their fields, linked to one another and to universities and government agencies. They designed and staffed scientific laboratories in industry, in the government, and in universities, and developed graduate training programs. The early Victo-rian scientific world, consisting of overlapping communities, with practi-tioners able to move across fields at will, solidified and developed boundaries that are still with us. The community broke into amateur and professional camps, and while the lines between the two were indistinct throughout the nineteenth century, in the twentieth they would solidify and take their contemporary shape. Under the pressures of specialization and professionalization—pressures that are felt just as acutely today—

the intellectual landscape of Victorian science, consisting at first of natural philosophy and natural history, broke into the modern disciplines that we study today.

Even in the last years of Victoria's reign, the British scientific world was simultaneously far-flung and small, globe-girdling and intimate, expansive and hierarchical. London was as central to British science as it was to imperial administration and finance. The leaders of the scientific community were well known and collectively acknowledged: they presided over the societies, occupied professorships at Oxford and Cambridge, directed government facilities and committees of inquiry, and mediated between the worlds of science and state. Scientific communication and networks exchanging data, field samples, instruments, and photographs went through London. The Kew coordinated a traffic in seeds and plants, the British Museum was a clearing-house for insects and animals, the Astronomical Society orchestrated a ballet of photographs sent to and from observatories all over the world. The careers of many scientists described a trajectory that began with training at Cambridge or Edinburgh or South Kensington, climbed with fieldwork or teaching in the colonies, and culminated with a professorship in England.

British scientific expeditions reflect in miniature the expansive and centralized character of Victorian science. Organized in England, by lions of the scientific establishment working with colonial lieutenants; paid for by private institutions, the navy, or the Treasury; coordinated by scientific societies in London; and tracing a path from Euston Station and Southampton to "darkest Africa," the Caribbean, the Pacific, Asia, or the poles and back, scientific expeditions bore the marks of the worlds in which they were created as strongly as those that they studied and through which they traveled. How those expeditions operated in the field, what they saw, and what sense they made of their experiences, are the subjects of the next chapter.

CHAPTER 3

The Experience of Fieldwork

Stories from the Field

On a spring evening in London a century ago, a rapt audience at the Crystal Palace listened to a lecture celebrating British mastery of science, imperial greatness, and superiority over colonial peoples. It was the keynote address to the Royal Photographic Society's International Photographic Exhibition, delivered by William H. M. Christie, Esq., C.B., Astronomer Royal, and director of the Royal Observatory at Greenwich. The event was presided over by James Lindsay, F.R.S., F.R.A.S., president of the society, 26th earl of Crawford, ninth duke of Balcarres, and benefactor of the Royal Observatory of Edinburgh (which he endowed) and the British Museum.[1] Christie and the earl were long acquaintances, and perhaps friends. They had both been at Trinity College, Cambridge, in the late 1860s (Christie was Fourth Wrangler in 1868, Lindsay left without taking a degree), and they had both been involved in Royal Society and Royal Astronomical Society affairs for decades. The subject of Christie's talk was a solar eclipse expedition to India from which he had recently returned. He began his lecture by explaining the importance of eclipses to modern science and describing the photographic instruments used by astronomers to study the sun. He described his party's journey to India on the P&O steamer *Ballarat*, from Portsmouth to Bombay via Suez, and acknowledged the help of the Indian government, railway companies, and the Admiralty, all of which provided essential services to his and other scientific parties when they made camp in India. At this point the lights in the hall were dimmed. Christie showed a series of lantern slides of previous eclipses, his expedition's camp, their instruments, and other objects of interest.[2] The expedition had been a success, he concluded, but it had not been without its moments of anxiety. "The natives of India," Christie explained,

regarded it as a very solemn religious function . . . and we were somewhat nervous as to whether our native assistants—who were high-caste Brahmins—

might not fall down on their knees and begin to say their prayers, instead of attending to the duties they were told off to perform. I am very glad, however, to bear testimony that they performed their duties most satisfactorily and managed to reconcile duty and religion. . . . I mention this as one of the things that cause some anxiety to observers during an eclipse, and to show that native ideas and native prejudices must be reckoned with.[3]

Christie's lecture was equal parts scientific paper, observatory notebook, and adventure story, and it is representative of accounts of the two dozen eclipse expeditions undertaken by British astronomers between 1860 and 1914. It alerts us to the fact that eclipse expeditions were not just scientific events of great significance, but had important cultural and social dimensions as well. In an era in which the sun did not set over the British empire, its eclipse was an opportunity to engage in a set of practices that produced valuable astronomical knowledge, embodied and asserted European mastery of the natural world, proclaimed England's dominance over its colonial territories, and demonstrated European superiority over non-Western peoples and races.[4] This is how they were experienced and described by British astronomers and commentators. Eclipse expeditions also allow us to study the relationship between the conduct of science in the late nineteenth century and the rise of imperialism. Astrophysical fieldwork was powerfully affected by the expansion of European technological systems around the world, the cultures of travel and colonialism, and ideas about colonized peoples. These forces shaped every aspect of expeditionary life and work, from the creation of new opportunities for astrophysical fieldwork, to the experience of being in the field, to the construction of observing practices used during totality.

This chapter is about the experience of eclipse fieldwork, and about the ways those experiences were interpreted and communicated to fellow Britons. The study of scientific practice has been quite popular in the last few years, yielding a number of highly detailed studies of experiments and fieldwork. Much of this work bears a resemblance to the cultural anthropology of Clifford Geertz, particularly his method of "thick description," in which rituals are described and analyzed in fine detail to yield lessons about the cultures of which they are a part.[5] This chapter has that same debt, for it seeks to understand eclipse fieldwork as the product of a set of cultures. The practice of fieldwork reflected the technical demands of astronomy, but it also absorbed much from European colonial culture and the newly minted culture of tourism. It also follows the work of scholars like Paul Fussell, John Keegan, and

Inga Clendinnen in trying to reconstruct and communicate some of the texture of the experience, a sense of what fieldwork was like for the astronomers, their wives, assistants, and colonial colleagues. The link between cultural histories of infantry combat and history of science may seem tenuous, but that literature's attempt to make sense of events that by nature resist description makes them helpful models in studying the confused experience of observing eclipses.[6] Finally, I want to put the two together, to show how culture defined expectations about what fieldwork would be like, and served as a tool for making sense of field experiences.

This enterprise will proceed in several stages. My aim is to show how the experiences of going on and writing about eclipse expeditions were shaped by larger cultural and technological forces in the Victorian world.[7] More specifically, I argue that eclipse expeditions were powerfully affected by the practices and culture of late-nineteenth-century tourism, and that they were further affected by expectations surrounding the encounter of civilized and uncivilized peoples.[8] I begin by analyzing published accounts of expeditions. These are a rich source of data about fieldwork, shaped by their authors to suit the tastes of readers, the conventions of science and travel writing, and the structure of the publishing market. These documents are cultural performances whose origins must be understood before they can be properly used.[9] I then follow expeditionary parties on their journeys into the field, show how they prepared for the eclipse, repaired instruments, and trained assistants. Expeditions made use of institutions and technologies developed for tourists, colonial administrators, and armies, and came to absorb some of the qualities of those institutions. (They also made use of the "tools of empire," to use Daniel Headrick's evocative phrase, in more substantive ways: I will argue in Chapter 5 that the presence of technologies like the railroad and telegraph made astrophysical fieldwork possible, and were indispensable to Victorian observers and their instruments.) Finally, I examine the experience of totality, and compare accounts of European and indigenous reactions to the eclipse.

Expedition Accounts and the Victorian World of Letters

"The graphic and lively letter which we publish this morning . . . makes the reader a companion of the journey of the scientific band, and admits him almost as an eyewitness to the intensely interesting observations which were at once its object and its reward."[10]

This reconstruction of eclipse expeditions will draw on a mix of unpublished sources, including private correspondence, telegrams from the field, reports to review committees, and diaries, as well as articles published in popular journals and scientific papers. My reliance on these sources requires a brief, critical look at them, the better to exploit their riches and deal with their limitations. They also reward analysis because they embody the kinds of mixtures of scientific and nonscientific forces that we will see shaping life in the field, the design of instruments, and representational practices. In this case, accounts of expeditions were crafted to take advantage of Victorian audiences' taste for travel narratives, and self-consciously combined the voices and conventions of scientific articles, travel accounts, and adventure stories.

SCIENCE AND TRAVEL PUBLISHING

The speech by William Christie to the Royal Photographic Society in 1898 was similar to printed eclipse accounts in terms of its subject, its language, and its messages. It was also shaped by the same literary conventions, market forces, and technologies that produced written accounts of eclipse expeditions. Eclipse observers and readers of eclipse narratives were part of the first generation to grow up consuming the products of the vast literary market that was one of the great innovations of Victorian society. Newspapers, monthly and quarterly reviews, and scientific and mechanics journals all fought to instruct the Victorian "articulate classes," while novels sought to entertain them. All pursued strategies combining high-volume production, large-scale marketing, and low cost—exploiting technological advances in printing and binding and catering to a growing, urban-based market of private readers and institutions.[11]

Eclipses were publicized and popularized by observers in newspapers, magazines, and books. Especially in the 1860s and 1870s, accounts of eclipse expeditions were published not only in scientific journals but also in popular magazines and reviews.[12] They were written to capture readers interested in science, travel, adventure, and the exotic. This literature existed at the intersection of two historical events: the emergence of the eclipse expedition as an ongoing scientific enterprise, and the rise of a mass, urban-based literate public. Months before astronomers packed up their instruments and set off for India or Algiers, long-lead articles (as they would be called today) would describe the importance of eclipse research, the state of scientific knowledge about the sun, and the plans of the researchers. The *Times* would cover the departures of researchers; *Nature* would follow their journeys to the observing sites and keep readers informed of the state of preparations. Once the eclipse had been observed, the same journals would reprint telegraphic messages received from the parties in anticipation of fuller accounts. Such accounts often extend over several weeks, as reports from foreign observers came in and scientists had time to analyze and pronounce on the newest findings.

These popular articles reflected the social attitudes and expectations of their readers. Science was presented as cumulative and constructive, the product of both disciplined teamwork and individual excellence. Eclipses were depicted as quasi-religious events. New discoveries were cause for nationalistic pride. Representative are the articles Oxford professor Rev. Charles Pritchard wrote for *Good Words* between 1867 and 1871. Pritchard hoped to convey "an adequate impression of the . . . strangeness, suddenness, awfulness, and majestic beauty" of an eclipse in his article, as well as describe its scientific importance.[13] He begins by recounting descriptions of the corona's appearance over the previous ten years. He then discusses the theoretical implications of different observations, the problems involved in properly drawing the corona, and accounts of the experiences of different observers. Most authors felt comfortable giving readers detailed descriptions of observational programs and research interests. Considerable space was given to descriptions of the voyages to and from the observing sites, escapes from potentially ruinous baggage-handlers, and the colorful reactions to the event by superstitious locals.

The gears of literary technology can be heard grinding in the background of this article. Lengthy descriptions of voyages, divisions of labors, and observations served to convince the reader that the reporter was a trustworthy eyewitness, attentive to detail and careful in description.[14] This literary technology was also tuned to the needs of popular publishing. Narrow escapes from Arab porters and clouds were highlighted to stimulate public interest in expeditions and increase demands for popular lectures and books on eclipses and astronomy. Eclipse observers were a remarkably literate group, and they included some of the period's most successful science writers and popularizers. Norman Lockyer and E. Walter Maunder, two of the most regular eclipse observers, were extremely active literary figures. Lockyer founded and edited *Nature*, explaining the journal's interest in eclipses. Maunder coedited *Observatory* with William Christie and contributed columns to *Nature* and *Knowledge*, founded by Lockyer nemesis Richard Proctor and edited by 1870 eclipse veteran Arthur Ranyard. Both Lockyer and Maunder wrote numerous popular books on solar physics, popular astronomy, and astronomy in the ancient and biblical worlds. May Crommelin, wife of Greenwich astronomer A. C. C. Crommelin, wrote more than thirty popular romance and travel books, Mrs. David Gill authored "an unscientific account of a scientific expedition" to Ascension Island, Winifred Lockyer translated French scientific texts for English-speaking audiences. The Lockyers' circle of friends included author Thomas Hughes, poet laureate Alfred Lord Tennyson, scientist Thomas Huxley, and publisher Alexander Macmillan, leading men of both the literary and scientific worlds. Other eclipse observers had connections to the world of popular science. Manchester professor Henry Roscoe organized popular evening lectures to entertain mill workers who were thrown out of work by the blockade of the Confederacy during the Civil War. The project was so popular that he continued it for decades and was able to enlist Huxley, Tyndall, Balfour Stewart, and other notables. This was a circle of people—Lockyer, Ranyard, Maunder, and their wives—aware of the ways that popular appetites for eclipse books could be whetted by *Nature* and *Living Age* articles highlighting the dramatic and romantic elements of the event. Their writing was finely tuned to the task of attracting and satisfying popular interests; and, to a large degree, they appear to have succeeded.[15]

OBSERVATORY NARRATIVES

Authors of eclipse narratives had another, slightly offbeat model: popular articles on observatories. Writers could have organized these articles, which described the history, layout, instruments, work, and national importance of scientific facilities, in any number of ways. But without fail, they chose to structure them as travel narratives.

Thus an 1872 article on Greenwich begins with a historical overview of the observatory, then moves readers into the present, with a quick walk around the grounds. "We must make the most of our time," the author says when the tour is concluded, "and enter the precincts of the Observatory." Once inside, the reader is guided from room to room, and their instrumental contents are described in detail. It is an exercise in virtual tourism. "We have spent a dreadfully long time in the Transit Room," the author says at one point, and hurries the reader, "with noiseless tread, through the Computing Room." From there, they "ascend to the top of the tower." After a quick look at the meteorological equipment and the London skyline, they "go down a flight of steps, into subterranean regions."[16]

One reason these descriptions were cast as visits is that actually getting into Greenwich was no mean feat. Casual visitors were refused admittance, for they would disrupt work and pose a danger to instruments. (In fact, other observatories, such as the Pukowa Observatory in Russia, were placed outside major cities to discourage visitors and to stimulate community and concentration among staff.)[17] One had to be an admirer of science with the connections to secure the necessary introductions, and readers were cast in the role of well-connected savants, scientific "insiders" in a real sense.[18]

Writers themselves played with this assumption of roles as a way of bringing readers into the story and turning virtual tourism into virtual witnessing. Edwin Durkin, a staff member at Greenwich who wrote on astronomy for several popular magazines (and who was involved in planning eclipse expeditions in the 1870s), published three articles in *Leisure Hour* that relied heavily on this conceit. "It is hoped that the reader will accompany us through the establishment, relying upon the information which we are enabled to give, which will afford him some compensation for the exclusive rules which it is necessary to enforce, for the preservation of order in the institution," the first article began. He then took his readers through the observatory, starting first with a quick view of London and Greenwich from the tower, then down into the apartments (a path used by writers since the 1850s). The reader's privileged if imaginary status is the key that unlocks the observatory's workrooms, and even earns extra indulgences. Upon entering "the more private portions of the establishment," Durkin and the reader "offer our apologies . . . but we have a carte blanche, and are therefore privileged." They visit the Astronomer Royal, then "walk quietly into the next room . . . [and] look over the shoulders of some of the assistants" as they carry out their duties.[19]

Two issues later, Durkin took readers back to the observatory at night, a journey even more interesting and more unlikely, for "according to the rules, [at night] no person, whatever his rank, can claim admission." However, "though no visitor is admitted after dark, we will exercise our official privilege of entrance . . . trusting to our acquaintance with the internal management." The reader doesn't look over the shoulders of assistants doing computations; now they're at their instruments making observations. At the transit instrument, Durkin and the reader stop to watch an assistant scrutinizing Jupiter. The great planet was a favorite object among amateur astronomers, and conveniently it is a popular subject in the Royal Observatory that evening: at another telescope the reader comes upon an astronomer "holding in his hand a representation of Jupiter, which he has just finished, whilst he is giving the planet a few parting glances to confirm the delineations, on which all the peculiar cloud-like belts are faithfully given."[20]

Eclipse Expeditions and Tourism

As we saw in the preceding chapter, months could be taken up wrangling with committees over plans, pleading with government agencies, and corresponding with foreign governments, travel agencies, and hotels. After all this preparatory work, parties must have often felt anxious relief finally to be at dockside, say good-byes to friends and well-wishers, and start off for the field. After a difficult journey from London to Portsmouth, for example, members of the 1860 expedition were restored by "the allotment of comfortable berths and a capital dinner" on their ship, the *Himalaya*.[21] Since eclipses almost always took place outside Europe and totality was so short, most of an expedition's time was spent traveling to and from the field, and setting up camp. Anywhere between several days and several weeks were spent traveling from one's home to observing stations. Some astronomers had trouble traveling by sea: for Rev. Stephen Perry, an astronomer who observed eclipses and transits and conducted geomagnetic research for thirty years, long sea voyages were a "martyrdom."[22] Astronomers traveling on the H.M.S. *Urgent* from Portsmouth to Oran in 1870 ran into several days' bad weather. John Tyndall recalled that during one evening "it was difficult to preserve the [dinner] plates and dishes from destruction"; he later awoke suddenly to find that "my body had become a kind of projectile, which had the ship's side for a target."[23] A pleasant voyage, in contrast, was a mixture of social event and informal scientific salon. Scientific travelers moved in an exclusive circle on passenger ships, dining with ship's officers and attracting curiosity and interest from regular passengers. Little work could be done on such ships, so scientists were obliged to entertain themselves and their shipmates. Aboard the P&O steamer *Mirzapore*, traveling from Southampton to India in 1871, Norman Lockyer gave informal lectures for passengers and crew on recent advances in science.[24] One member of the British expedition of 1870 wrote of his voyage from Naples to Sicily:

[D]elicious sunny afternoon, a sea as smooth as molten glass, a ship's company receiving us with the utmost kindness and hospitality, how could the evening not pass as merry as a marriage bell? I cannot retail to you all the jokes which passed, the lively chats and quiet strolls by moonlight, the polariscopes and spectroscopes pointed to the sea and sky, ere long destined to address their momentous questions to the Sun himself, now having their merits and demerits freely discussed by the savants; but you can imagine it all.[25]

Ironically, the next day his ship struck a rock and sank.

THE RISE OF TOURISM

Such mishaps aside, eclipse travel usually bore a strong resemblance to Victorian tourism, and in fact depended on many of its technologies and institutions. Eclipse expeditions were domesticated versions of the grand naturalist expeditions of the previous century, for they were conducted in lands controlled by European colonial powers and their allies, not in territory that was unfamiliar and potentially hostile. Disciplinary and instrumental constraints worked to confine the eclipse expedition to exotic but well-controlled ground, and the social rules governing behavior within that ground shaped the practices and experiences of expedition members.[26] Eclipse expeditions were planned and conducted in a social and technological environment radically different from that of the naturalists' expeditions of even fifty years before. The steamship and railroad had revolutionized travel, making new areas of the world accessible to large numbers of people, many of whom had never traveled great distances before. Technological developments promoted the rise of new institutions for serving tourists—hotels, travel agencies, and travel literature—and areas economically dependent on tourist money. In turn, social customs were created to guide behavior in these new institutions: how to travel, where to go, what to see, and how to feel about the experience were all laid out and understood by travelers.[27]

Infrastructure. The steamship and railroad changed the scale, speed, and character of travel. The steamship reduced voyage times between Old and New World ports to about ten days, and took much of the uncertainty out of transatlantic voyages.[28] After the opening of the Suez Canal, voyage times between England and Bombay fell from six to four weeks, and down to eighteen days by the end of the century.[29] The railroad made inland continents accessible to traffic and trade, and routinized travel between old routes; in so doing, it helped change passengers' perceptions of distance and time.[30] For those who could afford it, traveling became very comfortable: passenger liners were outfitted with lavish decor, posh dining facilities, and even decks for promenading, and first-class railroad service provided excellent food, comfortable accommodations, and impressive stations.[31] As a result, according to one travel agency, "for every person who traveled for pleasure fifty years ago, one thousand traveled in 1882."[32] One traveler summarized all these developments when he wrote: "Facilities in travel, more wonderful than the dreams of ancient poets, await the modern voyager, annihilating difficulties of time and circumstance, smoothing his path in the wildest regions, making his journey a mere question of time and money."[33]

Material changes in travel produced changes in the social organization and status of travel as well. The grand hotel evolved to accommodate large numbers of steam and rail travelers and provide accommodations conveniently near ports and stations.[34] Entire areas of Europe, including major cities, the Riviera, and the Swiss Alps, developed tourist economies, economies whose very existence depended on the new transportation systems.[35] Thomas Cook began offering package rail tours to temperance groups in the 1840s; business quickly expanded, and tours of London and the Crystal Palace, and then the Continent, were added in the 1850s. In the following decades, the company followed in the path of British colonial expansion and railroad extension, making travel to even Africa and Asia accessible to tourists by the 1870s.[36] In short, as travel changed in the late nineteenth century, it developed its own institutions and technologies, with the railroad and passenger liner, the grand hotel and resort, and packaged tour and guidebook all turning travel into a socially distinct and accessible form of middle- and upper-class recreation and education.

The culture of tourism. This tidy physical and social world had clearly established rules guiding action within it: where to go, how to prepare, and what to see and learn were all laid out in careful detail.[37] Different destinations would require different forms of preparation. The British traveled to Italy to see the products of a culture superior to their own, but they traveled to India to celebrate the triumphs and advances brought by imperialism to the subcontinent, and to see how missions and other European institutions fared under the Indian sun. Indigenous Indian architecture and art were treated by all but the most sensitive visitors not as remains of advanced and complex civilizations—the interpretation given to Italian artifacts—but as museum pieces and exotic curiosities.[38]

Regardless of destination, however, foreign travel had an edge of danger to it. Without the proper guidance and constraint, travelers leaving Britain and immersed in disconcertingly exotic, unusual, and more permissive lands (particularly in the Mediterranean and Asia) could be led easily into disaster. Stories of Europeans corrupted by too much time in foreign lands—of merchants and officers "gone native," or wives relaxing their morals under the influence of tropical atmospheres—showed just how bad things could become. Further, travel for its own sake was incomprehensible. Travel for religious purposes, for cultural enrichment, or for health was acceptable; traveling without a clear motive was not.[39]

To guard against the dangers of going abroad, travel had to be turned into a more intellectual, didactic, and socially remote activity.[40] Tour groups had everything planned for them, while books like Grant Allen's *The European Tour* and the Baedekers gave exhaustive instructions on how to do it yourself.[41] These books presented travel as a form of learning as much as recreation, a form in which "as in the regular classroom, no time was to be wasted with irrelevant material."[42] Like any classroom, homework was essential to success. Readers were urged to study the language and "history of the people, reading the best works descriptive of the country, [and] become familiar with its currency."[43] Guidebooks outlined which countries, cities, buildings, monuments, and art works to see and which to avoid, advising the tourist to visit "the typical features of each country, . . . and not waste his time on modern atrocities like Madame Tussaud's, the Crystal Palace, or even the city of Berlin."[44] Not only that, they recommended what *other* books should be read as well: after all, travel writer W. W. Nevin said, "[T]here is no greater saving in travel—saving of time, money, fatigue, temper, and opportunity—than that which is made in the procuring of good guide- or hand-books."[45] This was not an easy task—reading lists of thirty or forty books in two or three languages, most scholarly tomes and multivolume series, were not at all unusual—and it shows how seriously travel was taken in this period.[46]

Eclipse observers took their tourism duties as seriously as anyone. One writer, describing a visit to a botanical garden in Trinidad before the eclipse of 1889, wrote guiltily that "it was so oppressively sultry we could hardly crawl along, much less give our minds vigorously to the study of plants and names, and happily for our credit had we been questioned, many are without labels."[47] The conscientious traveler would thus arrange an itinerary based on advice from travel books, study the history and culture of the countries on his or her itinerary, visit those important places in the company of one's fellows under the guidance of Baedeker's or Murray's, appreciate their significance with the help of writers' commentaries, and return home a more refined and cultured person.

ECLIPSE FIELDWORK AS TOURISM

In this context, expeditionary travel does not appear all that different from tourism. In fact, the greatest difference between scientists and other travelers was that they were far more socially privileged (except for the aristocracy or very wealthy). In the Mediterranean, within the large British community of tourists, artists, writers, and scholars, archaeologists "had the mark of privilege," according to John Pemble. "Set apart not only by the arcane nature of the craft, but by their access to public money and official honours, they were the *crème de la crème* among students in the Mediterranean."[48]

Eclipse expedition members bore the same mark. Often eminent men of science, and on a mission of the highest scientific importance, they were treated as foreign dignitaries, dining with colonial officials, attending parties thrown in their honor, and touring gardens, universities, and libraries. John Tyndall was given various tours and dinners in Gibraltar in 1870, and later wrote in an autobiography that "our kind and courteous reception . . . is a thing to be remembered with pleasure."[49] British astronomers traveling in India for the eclipse of 1898 spent weeks "in camp" in the company of such colonial elites as the directors of the Indian Civil Service (ICS) and Great Trigonometric Survey.[50] The indistinct divisions between scientific travelers and others were matched by blurred distinctions in this period between "science" and other social activities. All amateur science was looked upon as a form of leisure, and its outdoor forms shared the same seasons as cricket and rugby. The sport of mountaineering even evolved out of scientific fieldwork.[51]

Finally, eclipse observers were a socially noteworthy group. The 1870 expedition, for example, looked like a page from *Who's Who*. Its members included half a dozen young Cambridge Wranglers, recent Oxford graduates, future MPs, eminent men of letters, the bar, and architects, a future earl, and members of the most prestigious clubs in London. These men were connected through talent, family, and marriage to what Noel Annan termed the Victorian intellectual aristocracy. Norman Lockyer came from a modestly successful provincial family, and was the son of an itinerant science teacher and lecturer. Many more came from prosperous and noteworthy families. Architect Francis Cranmer Penrose (eclipse of 1870) was a descendant of Edmund Cartwright, the inventor of the power loom; his cousin was Matthew Arnold. He later became a notable scholar of classical architecture and archaeology, surveyor of St. Paul's, and president of the Royal Institute of British Architects.

Other eclipse-watchers came from notable scientific families. Two of Charles Darwin's sons, George and Leonard, observed eclipses in 1870 and 1886, respectively. Brothers Charles Piazzi Smyth (eclipses of 1842 and 1851) and Warington Smyth (eclipse of 1870) were the sons of distinguished astronomer and hydrographer Admiral William Henry Smyth, and nephews and grandsons to a distinguished line of diplomats. Charles followed his father into astronomy, and eventually was named Astronomer Royal for Scotland, while Warington became a geologist. Their brother-in-law was Oxford mathematician the Rev. Baden Powell. James Herschel, son of Sir John Herschel, observed the eclipses of 1868 and 1871. John F. Maclear, who observed the eclipses of 1870 and 1871, was the son of South African astronomer Thomas Maclear. He married Julia Herschel, and had a distinguished career as a naval officer that included command of the oceanographic vessel H.M.S. *Challenger*.

While the composition of expeditions changed considerably from the 1880s, shrinking and becoming more professionally exclusive, they managed to retain that earlier flavor of polite society. Astronomer Royal William Henry Christie, a major figure in both expedition organization and eclipse observation from the 1880s to the 1900s, was the son of Samuel Hunter Christie, founder of the auction house. Edmond Grove-Hills was a graduate of Winchester and Woolwich, and sufficiently well off to pay his own way to the 1893 eclipse. But not everyone had such a rarefied background. Herbert Hall Turner's father was a portrait painter and photographer. At the tender age of nine, Herbert's occupation was listed in the Leeds census as "scholar," and he lived up to his early billing, winning scholarships to Clifton and Cambridge, and the Savilian professorship of Astronomy at Oxford.[52] (The amateur-oriented British Astronomical Association, in contrast, continued the tradition of volunteer-based expeditions into the 1900s, long after the RAS excluded amateur volunteers from its parties.)[53]

In sum, like all forms of scientific travel and fieldwork, eclipse expeditions developed their own sets of social conventions. Geological fieldwork in the first half of the century was ruled by a gentlemanly code demanding "zest, capacity for physical endurance, unfailing good humour, honesty, [and] bluffness"; eclipse fieldwork, in contrast, became an exemplar of privileged bourgeois travel.[54] Logistics, schedules, and itineraries were always worked out ahead of time, and little was left to chance. The social backgrounds of eclipse observers assured that they would be treated with deference and respect, and an expedition member's fellow travelers and hosts were people of the highest social standing and intellectual caliber. At least part—and often the entire duration—of an expedition would be spent in the same social world of ports, hotels, railroads, and government offices inhabited by Victorian tourists. Eclipse expeditions combined the didactic and recreational conventions of bourgeois tourism with the productive function of scientific fieldwork. Their organizers carefully researched the phenomena they hoped to see and made every attempt to ensure success in their observations; they enjoyed the advantages and comforts of the privileged classes on their travels, and were received as VIPs at their destinations; but they also produced new knowledge about the natural world at the end of their journey.

Life in the Field

Expedition members could look forward to plenty of attention at their destinations. Colonial governments often provided official aid to parties, and other Europeans volunteered their services. In 1871, the Indian Great Trigonometric Survey was ordered by the Home Department to provide the visiting British eclipse party with two assistants, 15,000 rials for expenses, "and all the aid that . . . [they] may require as regards photographic assistants, chemicals, & c."[55] Local officials, railroad engineers, and amateur scientists were an important source of aid to visiting observers. Isolated European engineers and administrators found the excitement a welcome change from their normal routine, and the astronomers appreciated the emergency repairs and supplies they could provide.[56] Volunteers, for their part, seem generally to have enjoyed the experience. Those who were amateur astronomers liked the chance to participate in serious science, and many expatriate missionaries and teachers obviously were gratified at the chance to share the company of European visitors.[57]

COLONIAL CULTURE AND ECLIPSE FIELDWORK

Colonial culture could also have its effect on eclipse expeditions, most powerfully in India. Three eclipses (in 1868, 1871, and 1898) were observed by parties sponsored by the Royal Society and RAS, and by Indian institutions such as the Royal Engineers and Royal Observatory at Madras. The main expedition in 1868 was organized by James Tennant, a member of the Royal Engineers; it was organized like a Trigonometric Survey party, with instruments borrowed from observatories and individuals in Britain and India.[58] Colonial administration of India contributed a set of conventions, artifacts, and culture built around fieldwork and the cult of outdoor life that were adopted by eclipse observers visiting from England. The ties between political power, Western culture, and scientific knowledge were drawn tighter and made more visible in India than anywhere else, because eclipse fieldwork in India owed less to other kinds of scientific fieldwork than it did to the culturally powerful and politically significant fieldwork of colonial administration.

Winter camping was an important part of British Indian official culture and the annual rhythm of work life.[59] Camping combined the duties and prerogatives of colonial life with the pleasures of travel and the outdoors. A typical camp consisted of an office tent, personal tents for officers and servants, and a dining tent. A camp's gear consisted of carpets, collapsible furniture, arms, a small library, and personal effects such as gramophones. Bullock carts or camels would carry the goods, and a half-dozen servants would cook, stand guard, make and break camp, and manage the goats and poultry. Favorite campsites were on the outskirts of villages, near a grove of trees and a well. From this site disputes would be resolved, taxes collected, and justice dispensed during the day. A few hours in the morning were reserved for hunting. But the winter months spent "in camp" were not designed to bring one closer to nature through primitive living. The entire ICS had only a thousand members, and one observer calculated that in 1900 about 1,200 Englishmen governed 232 million Indians.[60] Visits in camp were the only direct contact many Indians ever had with their government, and during those limited contacts camps had to communicate colonial power as effectively as government buildings in Bombay or New Delhi. They were official tours, moveable feasts of colonial custom, power, and privilege, and satisfied British perceptions of what impressed their Indian subjects.[61] One commentator wrote that camp administration "is government as they understand it in the East"; another, that Indians respect the "halo of pomp and circumstance" of official tours.[62] Even the location of camps on the edges of villages was given a historic justification.[63] Luxury and refinement were expressions of power, and camping a cultural form appropriated by British masters to naturalize and legitimate their rule.

Eclipse expeditions easily adapted to this system. Deputy commissioners often helped expeditions choose observing sites, with the result that, telescopes and other instruments aside, astronomers' camps looked exactly like those of district officials. A party of amateurs located in the village of Talni, for example, had a camp beside "a pretty grove of tamarind and mango trees, under the tallest of which the mess tent had been pitched," exactly as it would have been for an ICS official.[64] Once they were in the field, they were outfitted with tents and furniture from government stores, and helped to secure food and servants. The presence of volunteer officers from the ICS and Royal Engineers, who mixed rehearsals with hunting and formal dinners, completed the expedition's adoption of Indian camp culture. Those expeditions that were not outfitted by official organs were housed in a district commissioner's house or on official public grounds: James Tennant was hosted by the Guntoor tax collector and set up his instruments in the official's garden, and Edward Maunder's expedition of amateurs camped in the capital of the Buxar subdivision, at the local commissioner's invitation.[65] Two other parties stationed themselves in or near Ootacamund, a mountain resort and cinchona-growing area.[66] Another made camp among the coffee estates of Neddiwatam after being told that he "would be able to get European conveniences" there.[67]

Expeditions to the West Indies bore the marks of a different sort of colonial culture, less pomp and circumstance, and rather less drama. The well-developed tourist and travel-writing industries that churned out tourists to and books about India had no counterparts in the Caribbean. Those few visitors who did venture to the West Indies reported that the old jewels of the British empire were now tarnished and neglected. Norman Lockyer, Arthur Schuster, Edward Maunder, S. J. Perry, and Leonard Darwin arrived at St. George, the capital of Grenada, on 12 August 1886. In contrast to Bombay, which Lockyer and Schuster had both visited, St. George was a most uninspiring sight. James Anthony Froude, arriving at the end of the year, wrote that the town was a near ruin: despite its having one of the best natural ports in the region, sugar prices and commerce were at a low ebb, the dockyards had gone to the rats and weeds, and the town was "squalid and dilapidated." "Nature had been simply allowed by us to resume possession of the island," Froude wrote.[68] Roads on the islands were small and dusty, railroad lines were very limited, and travel was done on horse or on foot, or preferably by boat. Parties in the West Indies stayed at plantations and country houses, as the guests of local planters and colonial officials. There was no tradition in the West Indies of winters "in camp," since the islands were too small to require mobile administration, and the tourist industry in the 1880s was virtually nonexistent. A few mediocre hotels were to be found in the larger port towns, but almost no accommodations were available inland, and as late as 1903 there was no Baedeker's or Murray's *Handbook* for the islands.[69] Instead, visitors had to rely on the hospitality of private individuals, secured through letters of introduction and telegrams from friends of friends back in England. "Introductions to a few residents" were essential, one writer warned, if travelers wanted to be accepted into "pleasant society."[70] Darwin and Schuster were guests of the colonial treasurer, Turner and Lockyer set up camp at the home of a retired colonel near Grenville Bay, and Perry and Maunder were given a cottage at "The Hermitage," an estate on Carriacou Island.[71] Amateur astronomers traveling independently were likewise reliant on the hospitality of strangers and acquaintances, and found lodgings in estates and government residences.[72]

Expedition members often wrote of the difficulties they encountered "in the field" as a means of dramatizing and ennobling their experiences. In fact, while a few observing stations *were* located in dusty villages, those villages were often local centers of trade or government, where observers could find accommodations in villas and hotels. The experiences of an Australian party observing the eclipse of 1871 in the South Pacific, which encountered "legions of rats, who . . . boldly eyed the operations in the daytime, winking wickedly from behind the tufts of grass," and gnawed "at hats and any baggage which promised a toothsome morsel," were extreme.[73] Much more representative was the experience of H. H. Turner, who described the scene at his station in Grenada in 1886 in appropriately mild terms: "Anyone transplanted from home in his sleep and waking in an easy chair in the shady verandah at Prickly Point would be a minute or two before he discovered he was no longer in Europe—that is if the solitary date palm standing close in front of him were removed" and if the India-rubber bath were out of sight.[74] Another astronomer described Caroline Island, site of the 1883 eclipse, as "a most charming spot," with a lagoon that "was a never failing source of interest." However, he added, "we did not live as sumptuously as we did in Egypt," where they were guests of the Khedive.[75]

SETTING UP CAMP

Eclipse camps had a standard form, regardless of their location. Every camp had shelter for instruments and baggage, and a darkroom for developing photographic plates. Observers setting up in monasteries, villas, or estate gardens simply occupied existing buildings, while in unsettled areas, tents or huts were erected, and brick foundations and pedestals built.[76] Eclipse stations were also easily identified by the temporary observatories that parties built. These buildings were usually rectangular, and built of wood on a brick or stone foundation; they housed the larger and more valuable telescopes, spectroscopic cameras, and darkroom, and appeared everywhere between Iowa City and Africa.[77]

After securing or building accommodations for the party and instruments, observers would see to their instruments. At this point an expedition's tone would turn more serious, the frivolities of tourism replaced by the hard work of carving a working observatory out of the field. Telescopes and spectroscopes were unpacked and checked for damage, the camp's longitude and latitude were measured, and chronometers would be taken to telegraph offices to check their time against observatory time signals.[78] The process of unpacking, calibrating, and testing equipment was trying and time-consuming: instruments taken into the field always seemed to find new and unusual ways to break, and delays from bad weather, confusion over schedules, unfilled orders for supplies, and "Customs House officialism" were common. These were the problems of the lucky; less fortunate parties could face epidemics, small wars, and other unforeseen events. One expedition set out for a factory town in Angola only to find that it had long ago been abandoned and swallowed up by the jungle. Another spent two days idling in harbor in Brazil because they had arrived during Carnival and none of the longshoremen were coming to work.[79] "It was only through incessant work from morning till night" for two weeks, H. H. Turner recalled in 1886, that "all was ready . . . for the four minutes of darkness so precious to science."[80] After that experience, Turner warned others, "You cannot well have too much time [before totality]. It slips away very quickly."[81]

Efforts were also made to give the camp an identity as a workplace requiring the proper behavior of residents and visitors. Bunting or flags were sometimes placed over a camp's entrance to emphasize its separate identity, and guards were posted to keep the curious from playing with instruments, disturbing the astronomers, or even carrying away mementos.[82] De la Rue had five guards with him in 1860 to keep a couple hundred bystanders a safe distance away and quiet.[83] At the same eclipse, amateur William Pole left the town he originally chose after seeing that the authorities had constructed a guarded viewing area "enclosed with ropes like a prize ring"; he headed for the hills.[84] Astronomers and their volunteers kept to the camp until the eclipse was over, mimicking the social and physical isolation that was a hallmark of life at many observatories.[85] Expedition members visited and were entertained by dignitaries while traveling to and from the field, but once there the pattern was reversed: parties dined and slept in camp, and local officials and notables called on them. Casual visitors were kept out entirely, just as in regular observatories.[86] Once the camp and instruments were put in order, astronomers would teach local volunteers or soldiers how to use instruments or assist in their observations.[87] As instruments became more complex and astronomical knowledge more inaccessible, local elites became less important as independent observers; those who wished to work with visiting astronomers did so as assistants, operating instruments under the instructions of the professional peers.[88] Teams would then spend several days rehearsing their observing routines, making final adjustments to their instruments, and experimenting with new techniques and photographic developers.[89] The temptation to tinker with instruments after they were properly adjusted prompted one photographer to instruct observers to "accurately focus the sun a day or two before that of the eclipse, then remove the key of the focussing screw and *lose it*."[90]

LOCAL CONTACTS

All this activity inevitably attracted the attention and curiosity of locals. Sometimes they expressed amazement or superstition at instruments. "The delight of the Somali boys at being photographed was a sight to see," Lockyer wrote, "their broad grins being in strange contrast with the evident anxiety of the Arabs among the crew to escape the influence of such a possible evil eye."[91] More often their visits were occasion for exchanges that—for astronomers and their audiences— served as embodiments in miniature of colonial difference. American astronomer Eben J. Loomis described his encounter with African tribesmen as an object lesson in the difference between savage and civilized men, and the inability of the former to understand the latter. They could not communicate verbally, but Loomis had no trouble deciphering the Africans' pantomime: it meant, he wrote confidently, "[G]ive me rum." (This ability to make sense of native communication was a talent many Europeans seemed to possess.)[92] Still, the two sides looked at each other across a huge divide of civilization, "the accumulated effect of a hundred centuries of [European and American] effort . . . to understand and use natural laws for the benefit and uplifting of the race." Reflecting on what they made of their visits to camp, Loomis thought, "It is not at all probable that one of the dusky lookers-on at our preparations had a remote idea of the approaching phenomenon, and certainly not of the objects of our arrangements. . . . No effort could have given them the slightest comprehension of the causes of the unusual darkness, nor why the white man should come so far to look at it."[93] Notice in particular the way instruments highlight the difference between African and American: they were "certainly not" comprehensible to his visitors, while Loomis himself can take a double pride in both understanding them and serving them. After all, he has crossed the oceans to set them up in that particular spot. For the natives, in contrast, instruments, natural phenomena, and the behavior of whites were all equally inscrutable.

Astronomers and their instruments *did* attract the attention of natives, but for reasons that Loomis never fathomed. To local observers, it seems to me, *astronomers* were the exotics. Strange figures possessed by feverish obsessions with bizarre machines, astronomers sent out mixed social signals, combining access to colonial power with flagrant violation of colonial norms. The crowds gathered to watch men who were highly educated and well connected to powerful figures (figures who most of the time were just important names, known vaguely but not directly, and therefore all the more impressive), but who also behaved like coolies or madmen. "It never fails to puzzle" Indian onlookers, Elizabeth Campbell wrote, "to see the astronomers working for weeks like so many day-laborers (only much more rapidly) and then in the days before the eclipse commanding their forces like generals."[94] The sight of scientists pouring concrete, building piers, setting up instruments, and fine-tuning practices was at best odd, especially in places where Europeans were almost never seen doing manual labor. A British archaeologist in India remarked that Americans "astonished the natives, for there was nothing they could not do, and did not do, with their own hands."[95] Elizabeth Campbell said the same thing of Indians' reactions to her husband: "On-lookers . . . said they never saw a white man before who did all the work himself instead of ordering it done," she wrote. Soon, in fact, they were telling stories about him: "He is working from before dawn till after the sun has left the sky. Stones that four men cannot move he lifts with ease. And he is *never* tired!"[96] But this was nothing compared with what came next: this eccentric white visitor who could lift stones with ease would direct civil servants, a commander of the Bombay Marine, Royal Navy officers, and missionaries at rehearsal, showing them how to make observations, correcting their movement, ordering them to try again, chastising and encouraging them like schoolchildren. Likewise, an expedition in Labrador had assistance from the manager of the Hudson's Bay Company, a well-known medical missionary, a British Navy commodore, and the governor of Newfoundland. This caused consternation among the local fisherman. "When they first came, they [astronomers] wore rough clothing and did all manner of work like anybody else," an astronomer reported them saying. But in the days before the eclipse, "They bossed Mr. Swaffield! They bossed Dr. Grenfell! The bossed the Commodore! And they bossed the Governor!!"[97] This was something natives had never seen before, and probably didn't expect to ever see again. Small wonder they gathered to watch.

PRESSURE

The final days and nights before the eclipse were always tense affairs, as energies wore down and nerves grew strained. Lick astronomer John Schaeberle always had nightmares the night before an eclipse.[98] James Tennant complained from the field in 1868: "The climate of this place has been of great anxiety to me. . . . I have had many delays and troubles and have suffered so severely mentally and bodily that I do not feel myself. . . . I must just hope that I shall not break down" before the eclipse.[99] Manchester physicist Arthur Schuster, barely twenty-four when he headed an expedition to Siam in 1875, recalled that in the days before the eclipse, things went so badly that "I could not help sitting down and having a good cry."[100] (A vacation in the Indian resort of Simla and a tour of Kashmir restored Schuster's nerves.) Bad weather could put even greater strain on nerves. One astronomer recalled listening to a thunderstorm the night before an eclipse "with feelings akin to despair."[101] Others responded to the impossible expectations and pressures with dark humor. One lady volunteer, breathlessly relating her duties in a short-handed party, epitomized this response:

"I have to get three slides in and out," she told her traveling companion, "which have to be wrapped in black bags. I have to open shutters, to run round to another camera at the other side. One slide has to be exposed a minute, another half-a-minute; and this leaves me just half-a-minute to do all the manipulations. Then I have to shout the time for some *man* to put down, and he has in turn to shout back in *French*. . . . To crown all, I have to stand and jump on and off a pile of rickety packing-cases, and, *whatever* I do, I am not to give the telescope the *slightest* jar! I told Father Perry I was sure to come down and bring the whole apparatus down with me!"[102]

Despite the pressures, however, this work would not be *too* rigidly controlled: the eclipse party's organization still resembled the tour group more than the observatory, and strict routines and discipline would have been out of place in the field. Experience and technical knowledge were given value, but they only gently divided divisions of labor in the party. Generally, by the eve of the eclipse rehearsal experience mattered more than academic credentials.[103] Work routines could not be ordered too tightly, for as Greenwich astronomer E. W. Maunder noted, most members of a party were not astronomical laborers used to strict observational routines or "naval officers well accustomed to strict discipline and exact obedience, but independent ladies and gentlemen out on a holiday excursion."[104] Maunder endured strict, factory-like discipline at Greenwich in the name of moral order and mastery over the "personal equation," but in an eclipse party his experiences would be unique.[105] Many of his companions owned their own private observatories, where they were their own masters.[106] The backgrounds and training of most expedition members, therefore, made them unfamiliar with and unappreciative of the observatory's usual discipline.

The ease with which eclipse observers moved into and within the field is made clearer by comparisons with more famous expeditions of the era. Eclipse observers spent less time in the field, and encountered fewer hardships, than virtually any other group of scientists or explorers. The starkest contrasts are with the better-known geographic expeditions of the Victorian era. The Franklin Expedition, organized to search for the Northwest Passage connecting the Atlantic and Pacific, met an especially gruesome and famous end. Its ships, the *Erbeus* and *Terror*, were specially refitted for polar exploration, but even so they became trapped in ice in 1846 after a year of exploring. Two years later the entire crew was dead, perished after abandoning the ships in early 1848. Expeditions to Africa and Asia faced different but equally deadly dangers. The 1856–1859 Burton-Speke expedition, organized by Sir Richard Francis Burton to find the source of the Nile, sustained an immense amount of daily hardship and violence. The expedition was delayed after an attack in which Burton's palate was split by a spear. Once in the field, the two constantly battled fever and other sicknesses while porters deserted, hostile villagers attacked them, and animals and disease cut into their numbers. In this condition they lived for nearly two years without ever finding the source of the Nile. Their experiences were not unusual: one scholar has calculated that the average life-span of African explorers was forty-seven. In India at the same time, Alexander Gardnier, the first European explorer of the western Himalayas, survived starvation, ambushes, and attacks by wolves; George Hayward, explorer of northern Kashmir, endured similar hardships only to be killed in prison. One astronomer, the aged Rev. Stephen J. Perry, succumbed to dysentery while in the field in 1889, but that was a common enough ailment to befall European travelers in the tropics. Death simply was not the companion for astronomers the way it was for mountaineers or explorers. They never faced situations in which a single miscalculation could be fatal, nor did they have to make the wrenching choice of leaving behind ill or injured colleagues to save themselves, or risking their own lives to help their comrades.[107]

Expeditions that were not extremely dangerous could stretch on for years. Naturalists' expeditions were especially long. Henry Walter Bates spent eleven years in the Amazon, and Alfred Russell Wallace spent eight years in the Malay Archipelago. Surveying and oceanographic expeditions could also take years. The H.M.S. *Beagle* spent five years surveying the Latin American coastline and making a series of global chronometrical measurements. The American Wilkes Expeditions lasted four years, from 1838 to 1842.[108] The *Challenger* oceanographic expedition also lasted four years. Eclipse parties were not burdened with the hardships that came from such extended periods of fieldwork. In contrast to the rigors of mountaineering or surveying, eclipse expeditions were short, comfortable, and safe. Astronomers could spend half a year away from their observatories only if their travels included a long vacation or tour of Continental scientific institutions.[109] Observers suffered from little more than fevers, headaches, minor digestive problems, and seasickness, all ailments familiar to modern travelers. Few expected to be deprived of decent food, lodgings, and alcohol, and in general they weren't disappointed.

The high level of comfort and safety, no less than the social organization and behavior observed at these stations, remind us that they operated within the pale of industrial civilization, technology, and services. These stations were located in areas accessible by train and boat, and serviced by telegraph; the nearby railroad station might be small and dusty, but it *would* be nearby. It could hardly have been otherwise when expedition planners sought out colonial officials and missionaries to recommend areas offering easy access, telegraphic services, skilled assistants, and even good hotel accommodations. Accounts emphasizing stations' remoteness and exoticism must be read in perspective: Shirakawa, Japan, and Rawlins, Montana, might have been new to astronomers and gentlemen from New York and London, but they were never unfamiliar to the civil servants and European railroad managers who recommended them.

Work, Wonder, and Danger during Totality

THE ECLIPSE APPROACHES

Reporters organized their accounts of the eclipse itself around four stages. They began with the moon's first contact with the sun's disc. The two most important stages were second contact, when the moon completely obscured the sun and totality began, and third contact, when totality ended. The fourth stage was the end of the eclipse. First contact was a dramatic moment, but not a scientifically important one. Many astronomers didn't even observe it, and it mainly gave observers a chance to test their instruments and make last-minute adjustments or repairs.[110] As the moon moved over the sun's face, contact with large sunspots would be noted by those with telescopes. An astronomer looking at the image produced by a slit spectroscope aimed at the sun would have seen a bright band of light bisected by a number of fine dark lines. Shortly before totality, it would grow darker, the temperature would drop, and the moon would seem to move faster across the face of the sun.[111]

Writers filled these minutes not just with observations and last-minutes checks of instruments but also with dark foreboding and nervousness. Clouds, especially if they were near the sun, heightened the tension. "The decisive moment approached nearer and nearer; with anxious care we contemplated the clouded sky, when suddenly, to our great joy, the clouds parted, and we beheld the sun, already partially eclipsed."[112] The excitement in the last seconds before totality, especially when the weather was still threatening, was almost painful in its intensity. "The excitement became very great as the crescent was reduced to a mere line," Henry Holiday wrote to his family from India in 1871; "I could with difficulty control myself so as to be fit for making a decent observation, and Mr. Whiteside told me afterwards that his heart was thumping violently."[113] Warren De la Rue watched the cloudy sky in Spain with "the most intense anxiety," and felt his "nerves were in the most feverish state of agitation."[114]

Adding to the general tension and strangeness were surreal, almost supernatural changes in the sky and landscape. "The clouds began to look dark and threatening, and appeared to lower down towards the earth, while the parts of blue sky gradually changed to a deep somber purple," one writer reported.[115] Others elaborated: the darkening sky and strange colors "gave a ghastly illumination to the landscape,"[116] and cast clouds and stations in an "ashy gray colour."[117] People were likewise transformed by the changes wrought by the eclipse, casting on people an "unearthly cadaverous aspect,"[118] making them look like "denizens of another world, so livid did their faces appear."[119] The narrowing crescent sun would throw strange shadows on the ground, and bands of light and dark would play across walls and the sides of tents. These brought on "a sort of giddiness in the vision," and in the unprepared had an almost hallucinatory effect: an astronomer camped among Chilean miners reported that they saw "a strange appearance under their feet, which they felt quite persuaded arose from masses of green snakes."[120] The theme of earthly transformation accompanying cosmic transformation, of the world turning strange as the sun's atmosphere was revealed, was prefigured in pre-eclipse visions: spectacular sunsets, the revelation of a first look through a large telescope, the magnificent glint of a distant city's towers viewed from a train or steamer, were constant subjects of notice and description. By the time of the eclipse astronomers were prepared for wonder.[121]

SECOND CONTACT

Failure. Sometimes the clouds did not clear in time, and astronomers were denied the prize they had worked so hard for. Failure was a bitter experience after months of planning and weeks of work in the field. Clouds and rain often started hours before the eclipse began, putting observers in an anxious mood. Instruments were covered with tarpaulins to protect them against rain, and were uncovered the second it stopped. Often nature would throw out a bit of false hope—a bit of blue sky suddenly visible near the sun, a crescent sun breaking through the clouds for a few seconds—as if to cruelly emphasize the powerlessness of observers, and tarpaulins would be put up and taken down over and over. A clouded totality was dark and glum, but observers never left their posts, even when it looked completely hopeless.[122] "One had a strange feeling," A. A. Common said, "that something had gone wrong, for the darkness and the silence were very oppressive."[123]

Few bothered to hide their intense disappointment immediately after totality. "As the light returned it showed a very disconsolate group," one amateur wrote of the moments after a clouded-out eclipse. "We could hardly look in each other's faces, and I am sure no one just then could find voice to speak." It seemed, her companion told her, "like a death-bed, the more so as Dr. C. and his helper began at once to encase some of the instruments in the black, coffin-like boxes."[124] Foul tempers threw clouds on posteclipse sightseeing and could survive the return trip home. The *Daily Telegraph* reported after the failure of the 1896 eclipse that "the word 'eclipse' is tabooed under the penalty of a severe fine" aboard the ship carrying observers, and "astronomers now walk on the side of the ship away from the sun."[125] Some undoubtedly were mad at themselves: they had decided to cluster in a tiny area along the Finnish coast rather than spread out along the shadow path, as was the custom. It hardly needs to be said that the eclipse was visible only a few miles away.

The start of the eclipse. Almost no astronomer was denied success if he persevered and kept going into the field. Eventually he would find himself under clear skies, witness to the onset of totality, as dramatic a natural spectacle as any person was ever likely to see. The most dramatic moment was the passage of the moon's shadow over the camp. It had an almost physical quality, like the edge of a tornado or onrushing storm.[126] "[We] saw, or as it were *felt*, the mighty rush of gloom which came sweeping at an awful speed," Charles Pritchard wrote about the eclipse of 1860, "like a storm over the waters, and yet suddenly wrapping objects in an unexpected, windless silence and calm."[127] Another member of the British expedition described the shadow as "the most awful thing I ever saw . . . enveloping part of the sky in a dense shroud of the most fearful gloom."[128] Harvard College Observatory director Edward Pickering reported seeing "a veil of darkness rise from the horizon" and rush toward his party "with the speed, not of the wind, but literally of a cannon-ball."[129] A few miles from Pickering, an English amateur described it "producing the effect of the clouds having suddenly contracted and approached the earth."[130] As the shadow rushed over the land, those watching through spectroscopes would see some of the spectral lines grow brighter, then a bright flash—a reversal of the lines—and then a slightly different spectrum, with some lines removed and others brightened. A ring of bright beads would appear for a few seconds on the eastern side of the sun, then disappear. The observers were now inside a shadow, cast by the moon, about one hundred miles wide and traveling at two thousand miles an hour. Totality was upon the station.

It was now suddenly deathly quiet, and the world seemed more normal. "Up to the instant of totality," architect R. F. Chisholm wrote from India, "the motion of the heavenly bodies had been so rapid that fancy almost created a sound; you fancied you heard the whirl of the moon rushing on through space . . . and now, all seems perfectly still!"[131] The most striking and anticipated aspect of totality was the appearance of the corona. "As every eye watches eagerly the small glitter and dazzle of expiring sunlight," a journalist describing the eclipse of 1869 wrote, "it is suddenly transformed into an indescribably beautiful halo . . . a white radiating glory . . . bearing a striking resemblance to the light which painters draw around the heads of saints."[132] This unification of visual with religious imagery—the comparison of the corona to a saint's halo—was common.[133]

Others used images of wealth and political power. "There, rigid in the heavens, was what struck everybody as a decoration, one that Emperors might fight for; a thousand times more brilliant even than the Star of India," wrote a British observer.[134] Francis Galton said it reminded him "of some brilliant decoration or order, made of diamonds and exquisitely designed."[135] The corona's complexity and beauty inspired brighter emotions than the darkening skies and hurtling shadows that preceded it. "No feeling of terrified awe came over us" during totality, one observer wrote, "but a grander emotion was experienced, that of superlative sublimity."[136] Norman Lockyer, at his first successful eclipse, was surprised that "there were no ghastly effects" during totality, "no yellow clouds, no seas of blood . . . no death-shadow cast on the faces of men."[137]

Accounts in the 1850s and 1860s admit that the emotional impact of eclipses made both objective observation and reporting difficult. Charles Pritchard wrote that "conveying an adequate impression of the . . . strangeness, suddenness, awfulness, and majestic beauty" of totality was made difficult not just by the phenomenon itself; any observer who "has any sensibility to the sublime in nature" would naturally be "overpowered by the novel and supernatural effect of the scene."[138] George Airy warned that "the phenomena [of totality] are so striking that the most perfect discipline will predictably fail."[139] Warren De la Rue's dis-

cipline almost did fail. "I was so completely enthralled" at the view of
the corona, he remembered, "that I had to exercise the utmost self-control to tear myself away from a scene at once so impressive and magnificent."[140] "Certain it is, that no man of ordinary feelings and human
heart and soul, can resist it," Edinburgh Astronomer Royal Charles Piazzi Smyth wrote after the eclipse of 1851, adding that "it was not only
the volatile Frenchman who was carried away in the impulses of the
moment . . . but the same was the case with the staid Englishman and
the stolid German."[141] However, using the language of the sublime and
speaking of the emotional impact of totality in such frank, confessional
terms raised questions about the credibility of observers. Sensitivity to
the sublime was one thing, senseless shock quite another.[142] Piazzi
Smyth, while respected for his careful observatory work, was turned
down as an eclipse observer in 1858 because it was felt that he "is too
excitable in imagination and could not refrain from landscape painting."[143]

This language recalled the British tourists of the previous century
who had organized their travels as tours of sublime and picturesque
sights, but the revival did not last long.[144] The greater reliance on large
instruments and cameras for observation profoundly affected the character of observers, demanding attention to technical detail and operations. It also changed the structure of published articles: from the late
1870s, they devoted ever-larger amounts of space to describing instruments, observing procedures, and developing methods; accounts of
travel and camp life, which previously had taken up considerable
amounts of space, were reduced to a few quick facts, and accounts of
the onset of totality were stripped down to a few sentences, then finally
to the times of second and third contact.[145] By the end of the century,
dramatic effects before and during totality were notable for their absence in eclipse accounts: nearly everyone remarked in 1898 that the
onset of totality passed without lurid effects or sweeping, supernatural
shadows.[146] The language of the sublime even disappeared from unpublished accounts of the eclipse. Ralph Copeland, Royal Astronomer for
Scotland, described the landscape before totality as similar to "a thunderstorm in the height of summer."[147] Another wrote that while she
"had somehow expected, from lurid accounts of eclipses, some wonderful red, orange and blue fireworks filling the sky . . . the whole effect
was not grand nor terrible but exquisitely beautiful." Still, she added, it
was "so totally unlike anything I had ever seen or imagined that I felt I
must be in another world."[148]

Third contact. The end of totality came too quickly for those still working to finish observations or take photographs as the sun's disc burst forth; still, it was a signal relief, a sign that the world was returning to normal. "Like a flash of silver lightning, the first ray of sunlight shot suddenly out, piercing the gloom for a second, again illuminating the world . . . [in] a vivid conception of the Creator's grand dictum, 'Let there be light, and there was light.' "[149] Arthur Schuster reported, "When the sun burst out once more, a feeling of relief came over me that all the excitement was over, and at the same time a feeling of intense disappointment, a kind of presentiment of failure, due no doubt to the sudden relaxation of the nervous strain." He also reported that American astronomer Charles Young "once told me that a peculiar feeling always came over him within the few minutes following totality, which he did not have on any other occasion, and which he could only describe as the sensation of realizing that he was a mean dog."[150]

Indigenous reactions. The sublime or scientific responses of European observers to the eclipse were always contrasted with the primitive and suspicious behavior of local peasant or indigenous observers. Indians and Africans of all classes were depicted in a bad light, but European peasants were also targets of writers' attacks, especially when reports of their behavior reinforced nationalist or religious prejudices. Charles Piazzi Smyth described the effects of totality on Norwegian villagers in 1851: with the coming of totality there "suddenly came into view lurid lights and forms" that "made the Norse peasants about us fly with precipitation, and hide themselves for their lives."[151] Russian peasants crossed themselves and chanted prayers under direction of Orthodox priests; Spanish peasants conducted themselves in a similar manner.[152]

The strongest words were reserved for non-Western indigenes. African explorers recounted their observations of one tribe in 1830: "As the eclipse increased they became more terrified. All ran in great distress . . . [for] they could not comprehend the nature or meaning of an eclipse."[153] During the American eclipse of 1878, one observer reported that while he considered the eclipse "the grandest sight I ever beheld . . . it scared the Indians badly. Some of them threw themselves upon their knees and invoked the Divine blessing; others flung themselves flat on the ground, face downward; others cried and yelled in frantic excitement and terror."[154] "Even John Chinaman was expecting the eclipse" of 1878 in Denver, but rather than sketch the corona, the "almond-eyed celestials . . . beat their gongs all through totality" to drive away dragons.[155] Eight years later, H. H. Turner reported: "Much apprehension had been felt by the darker parts of the population of Grenada," caused by a rumor that "the island would be visited by a great tidal wave."[156]

The astronomers even reported themselves worked into some superstitions. This happened regularly abroad in India, Africa, and the Indies, but also in Catholic Europe. The (Anglican) Rev. Charles Pritchard wrote that Spanish peasants "supposed [his party] to be in league with the Evil One, from whom, for a consideration, we had purchased our knowledge of the eclipse."[157] According to Alfred Brothers in 1870, Sicilian peasants believed that the eclipse was divinely ordained "because the Pope had been deprived of his temporal power." More troubling was a second rumor that the foreign (and Protestant) observers "kept the seeds of cholera" in their instrument cases and were going to start an epidemic.[158] Russian Orthodox priests spread rumors that the "Anti-Christ had come over from America" with American astronomer Charles Young to "darken the sun" and hasten "the resurrection of the dead and the end of the world." Fortunately, Young's host was a government official with the stature "to counteract the influence of the priests."[159] Edward J. Stone, observing in South Africa in 1874, reported that "the natives . . . got up a tale that I had brought the eclipse with me."[160] In politically turbulent Egypt in 1882, Lockyer fretted when he heard that "a whisper had gone abroad that the False Prophet of the Sudan had included the eclipsers in his anathemas," and he ordered the military guard around his camp strengthened in response.[161]

India was the richest source of stories about native reactions. John Maclear, son of the Astronomer Royal at the Cape of Good Hope and a distinguished naval officer and man of science (he commanded the H.M.S. *Challenger*), printed several rumors that he claimed circulated "among the natives" after his party made camp in an old coastal fort. The first was that "part of the sun was about to fall, and the wise men had come to the East to prevent it. Then when the formidable-looking instruments were seen mounted on the fort, they thought there was a war." Finally, another rumor spread that "a flood was about to descend, and all the Europeans were coming to the high ground to escape it."[162] In Jaffna, Norman Lockyer was told that "the crowd of natives . . . were under the impression that the whole of the Expedition were going to get into a balloon and [fly] off to the sun."[163] A European expatriate reported that Brahmin priests turned the eclipse into an opportunity to extort money from villagers. The experience, he felt, showed the futility of modernizing efforts in India: "[M]any well-informed and educated natives performed all the superstitious ceremonies connected with the eclipse, with just as much zeal as the ignorant ryot."[164] In 1898, astronomers heard some Indians claim that since the establishment of the Raj the number of eclipses had increased "on account of increase of sins and misdeeds" by the British.[165] These stories had two effects. First, they showed that natives were not just prisoners of ancient superstitions but were able to create their own contemporary superstitions. Second, they served to flatter European narrators by turning them, however briefly and from however untrustworthy a source, into masters of nature.

But whether right or wrong, the morals to be drawn from these accounts, with their stark contrast between the behavior of Western and non-Western observers, and stories of opportunism and rumor-mongering of indigenous elites, are obvious. Eclipses are events that can be understood rationally, yet Orthodox priests and Brahmins exploit fears among the peasantry for their own purposes. On a personal level, Westerners always behave as masters of themselves and nature, non-Westerners as superstitious savages, unable to restrain themselves or understand the natural world. Again, Norman Lockyer expressed the contrast most clearly: "There is strict silence in the fort" where his party observed the Indian eclipse of 1871, he wrote, "and the work of recording ... the phenomena visible in telescope, spectroscope, and polariscope, before totality, goes on like clockwork; but it is very different below. The natives see in the eclipse their favorite god devoured by the monster Rahoo, and ... yells, moans, and hideous lamentations rend the air as the monster seems to them to get the upper hand."[166] There were occasional modulations in the tone of these descriptions—a chapter originally entitled "Eclipse Customs amongst Barbarous Nations" might be softened to "Strange Eclipse Customs," or a writer might note that Europeans had, centuries ago, reacted in similar ways—but their basic message stayed the same: modern European and non-Western reactions to eclipses were fundamentally different.[167]

However, natives could easily move from amusing or irritating to threatening. Lockyer's expedition of 1871 would have been ruined by "the smoke of [Hindu] sacrificial fires ... , if there had not been a strong force of military and police present to extinguish them; and in Egypt, in 1882, without the protection of soldiers, a crowd of Egyptians would have invaded the camp."[168] There was a world of difference between feeling sublime terror and awe at the terrible beauty of the corona and prominences—a reaction that, first and foremost, required an education in Romantic aesthetics—and being driven into a murderous frenzy by superstitious terror. Colonial masters by the 1860s had learned to equate superstition with fanaticism, and hence natives, reacting (according to European accounts) to eclipses as primitives, were dangers against which defenses had to be prepared.[169] Before an eclipse crowds were merely troublesome, but during an eclipse they were far more dangerous: stirred up by jealous priests, shackled by ancient superstitions, constitutionally incapable of the same kinds of self-control on which Europeans prided themselves, they threatened to revert to savagery under the enormous emotional pressures of totality. Army or police detachments, accustomed to involvement in "punitive [military] expeditions," were not just emblems of power or courtesies bestowed by local officials: astronomers saw them as absolutely essential field-workers.[170]

The most that can be said for these stories is that they reveal much more about European anxieties than actual native behavior; at worst, they are completely wrong. They probably drew on the sizable reservoirs of anxiety astronomers felt about the performance of native assistants. This is especially notable in India, where they played prominent roles as instrument operators and personal servants. William Christie worried that his Brahmin assistants would abandon their duties to pray, while Madras astronomer Norman Pogson opined that "rational natives are very scarce" under normal circumstances, and impossible to find during an eclipse.[171] They could also cause small problems. One of Warren De la Rue's Spanish assistants, charged with making smoked glass for bystanders, "became so excited" by the clamor for his product "that he threw away the matches in all directions without extinguishing them, and some, falling in the standing corn, set it on fire. Happily, a few seconds after the occurrence, the crackling sound and the smell of burning straw drew my attention to the spot, and, water being at hand, the fire was got under before it had spread more than a few feet."[172] Accounts of native reactions by nonscientists are spotty, but none suggest that better-educated and urban colonials ever thought that the sun was being devoured by gods or dragons. One Indian clerk told European astronomers that while his friends paid some attention to traditional rituals, they did so only to satisfy their parents; they themselves no more believed in the old ways than their British employers.[173] Novelist Violet

Jacob likewise saw the eclipse of 1898 at a lawn party thrown by a local Indian *Bankwallah,* and her diary, which is full of pointed observations about Indian character, says nothing about superstitious native reactions.[174] These accounts suggest that educated Indians treated an eclipse in precisely the same way most Europeans did, as an interesting scientific curiosity and excuse for a good party.

Conclusion

However, while these stories, and the published and unpublished accounts in which they appear, have precious little reliable information about native reactions to eclipses, they do tell us a great deal about British perceptions and self-perceptions. First, they show that expeditions blended hard science with tourism, and were conducted in spaces that were part observatory, part military camp. We can safely say that astronomers were in less danger than they believed, and dissect their reasons for exaggerating the threat of native violence. Colonial prejudice, beliefs about the comparative intelligence levels of Europeans and nonwhites, and chauvinism played a role, but so did something less reprehensible: astronomers saw expeditions as adventures, taking them to locations that were exotic and even perilous. Accounts of expeditions are equal parts scientific article, travel narrative, and adventure story, and this mix of influences reflects more than literary convention or calculation: it reflects the way astronomers thought about and experienced fieldwork. They wanted expeditions to be momentous and a little dangerous, and while the fact that no astronomer was ever felled by a savage mob speaks for itself, we need to recognize that that desire affected eclipse fieldworkers even as we note that the reality was different. Astronomers sometimes invoked the tradition of scientific travel exemplified by Capt. James Cook, who observed a transit of Venus in 1769, and declared themselves his spiritual descendants. In reality, the foundation, character, and texture of their experiences made eclipse expeditions less like James Cook's voyages than Thomas Cook's tours.

These sources also tell us something else. Carrying instruments into the field, getting them and assistants up and running, solving problems before an eclipse, and observing the eclipse itself were enormously hard work. No one should underestimate the challenges of astrophysical fieldwork, or consider the difficulty of the enterprise unimportant when trying to understand its history. That enterprise did not get easier with time: instruments grew larger, observing programs evolved, theories about the composition and structure of the solar corona changed, but the difficulties did not grow smaller. Even instruments that were designed specifically for eclipse fieldwork weren't easier to use or more rugged than their predecessors. As a result, while many other things changed, the basic challenges of doing solar physics in the field remained the same. We turn to those challenges next.

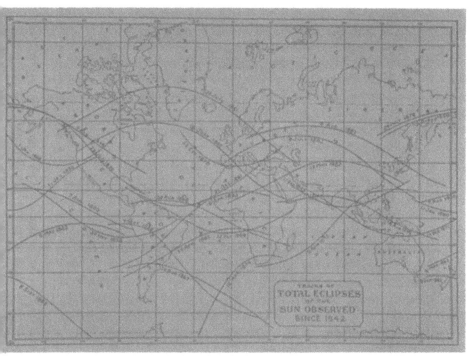

1. Total eclipses of the sun, 1842–1919. This map shows the paths of eclipses from 1842 to 1919, the year in which astronomers used observations of an eclipse to confirm Einstein's theory of relativity. Few eclipses occurred over Europe, making it necessary for astronomers to travel the world in pursuit of totality. (Samuel A. Mitchell, *Eclipses of the sun* [New York: Columbia University Press, 1923].)

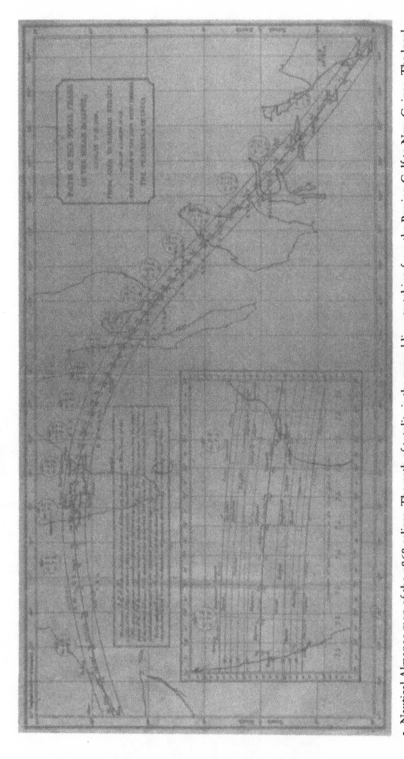

2. Nautical Almanac map of the 1868 eclipse. The path of totality is the curved line stretching from the Persian Gulf to New Guinea. The local time of first contact and duration of totality are marked along the path. The inset shows the eclipse's path over India and the locations of towns from which the eclipse will be visible. This smaller map was of particular interest to British observers: a number of British colonial government officers and expatriates organized eclipse parties. British observers would travel to India to observe eclipses in 1871 and 1898. (RGO 6/122, fol.

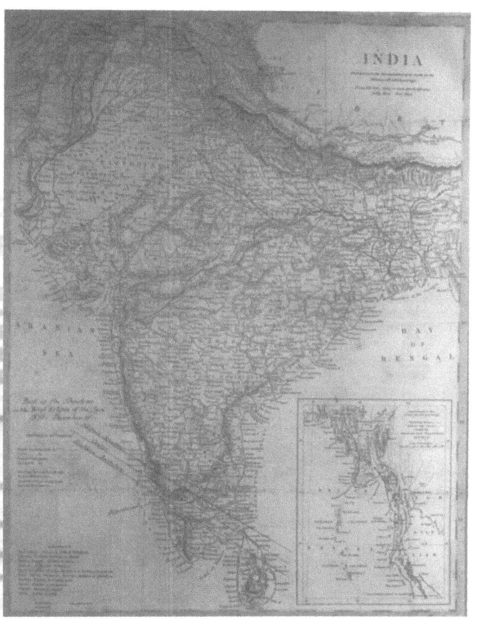

3. Map of India with path of the 1871 eclipse drawn in. (RGO 6/195. By permission of the syndics of Cambridge University and of the director of the Royal Greenwich Observatory.)

4. Spectroscopy at sunrise on the Indian Ocean. Travel to field sites could take weeks; indeed, the majority of an expedition's time could be spent in transit. Making scientific observations of shipboard phenomena was a popular way to pass the time, and for amateurs like this BAA party, may also have helped acquaint observers with their instruments. (From E. A. Maunder, *The Indian Eclipse, 1898* [London: Hazell, Watson and Viney, 1899].)

5. Eclipse camp, Sahdol, India, 1898. William Christie showed this picture to his audience at the Crystal Palace (see Chapter 3). This picture gives a sense of the improvised character of eclipse camps, with their temporary buildings, and sun- and wind-breaks made from grass. Note the three Indian workers on the left. Christie is on the other side of the instrument. (RGO 7/198, photograph 2. By permission of the syndics of Cambridge University and of the director of the Royal Greenwich Observatory.)

6. Another view of the eclipse camp at Sahdol, India, 1898. The white tents, loaned by the Indian Civil Service, were emblems of the imperial government's presence in remote areas. Winter camping was an important part of British Indian official culture; in addition to bringing colonial government attention to areas that otherwise would go overlooked, some British writers claimed that this was "government as they understand it in the East," and that Indians respected the "halo of pomp and circumstance" of official tours. An elephant stands in front of the second tent from the left. (RGO 7/198. By permission of the syndics of Cambridge University and of the director of the Royal Greenwich Observatory.)

7. British Navy seamen were an important source of volunteer labor for many expeditions; expatriate engineers, missionaries, and teachers were also recruited to bring expeditions up to strength. This photograph shows a group rehearsing before the 1896 expedition in Norway. (J. Norman Lockyer, *Recent and coming eclipses* [London: Macmillan and Co., 1898].)

8. Eclipse fieldwork in the 1870s. This engraving shows Norman Lockyer's party of amateur astronomers and gentlemen at drill before the eclipse of 1871 at Bekul, India. At the left and center are four telescopes on two mounts. Two observers would work together on these instruments, one making visual observations of a section of the corona or prominence, the other observing the object's spectra. This picture gives a good sense of the scale of eclipse observation in the 1860s and 1870s: each man works alone or with one assistant, none of the instruments are custom-made for eclipse work, and almost all observations are made visually. Note the packing crates used as instrument stands. (From J. Norman Lockyer, *Contributions to solar physics* [London: Macmillan, 1874], plate 6, facing p. 343.)

9. Eclipse drawing in 1871. This drawing by Henry Holiday shows himself drawing the corona at the eclipse of 1871: he observes the eclipse through one eye, and draws it with the other. Note that the telescope and paper are at nearly the same height; Holiday needs to move his eyes slightly to focus entirely on the drawing, rather than tilt his head down. Eclipse drawing was emotionally draining: As Holiday later wrote: "I could with difficulty control myself so as to be fit for making a decent observation"; just after totality, "[I] plunged my head into water, for I was in a fever of excitement." (From Henry Holiday, *Reminiscences of my life* (London: William Heinemann, 1914), fig. 13 on p. 209).

10. Large eclipse camera, India, 1898. This instrument was designed by Royal Edinburgh Observatory director Ralph Copeland (seated on the left). The mirror at the right directs the sun's image into the tube; the photographic plates are exposed in the darkroom on the far left. Copeland's is representative of the large, specialized instruments that became popular with JPEC expeditions in the 1890s, and raised the cost of eclipse observation, concentrated labor in the field, and produced more data than their predecessors in the 1860s and 1870s. (Courtesy of the Royal Observatory, Edinburgh.)

11. Large telescope and coelostat, 1898 eclipse camp. The eclipsed sun was reflected off a mirror at the top of the soil pyramid. A small darkroom is attached to the bottom of the telescope on the left. Three smaller telescopes are visible on the far left. As was often the practice, this camp was made in a cleared field, near a stand of trees (visible in the background). (RGO 7/195, photograph 13. By permission of the syndics of Cambridge University and of the director of the Royal Greenwich Observatory.)

12. A party of British Astronomical Association–sponsored amateur eclipse watchers in India, 1898. The small size of the instruments that BAA parties used contrasts sharply to the behemoths deployed by the professionally staffed, JPEC-sponsored expeditions. This photograph shows a "Mr. And Miss Bacon" rehearsing their observations for the eclipse; Miss Bacon keeps time with a metronome (visible at the left end of her instrument box). Note the party of Indian soldiers and workers in the background. (From E. A. Maunder, *The Indian Eclipse*.)

13. William Henry Wesley (1841–1922), artist and assistant secretary to the Royal Astronomical Society, in later life. Wesley began his career as a painter but shifted to scientific illustration in the 1860s. After working with the RAS on the *Memoirs* volume on recent eclipses, Wesley joined the RAS as assistant secretary, a position he held for the rest of his life. (From *The Observatory* 45 (1921), facing p. 341.)

14. Alfred Brothers's photograph of the solar corona taken during the 1870 eclipse in Spain. This is a positive copy made by Brothers himself; though it is nearly impossible to see in this reproduction, in the original one can see that the prominences are painted in. (Reproduced courtesy of Dartmouth College Library.)

15. A woodcut of the eclipse of 1868. Woodcuts were acceptable for use in newspapers and textbooks, but they could not sustain the level of detail necessary for real scientific work. The rendering of the streamers as groups of lines, and the mottling of the background sky, were unacceptably artifactual. Also notice that the prominences are made darker than the corona itself; the woodcut's ability to convey brightness was exhausted by the inner corona, so the engraver drew the prominences in reverse. This was an unusual, but not unheard-of, practice in astronomical illustration. (From Lockyer, *Contributions to solar physics*, fig. 113 on p. 289.)

16. Proof of a lithograph of the eclipse of 1871. This lithograph was a composite drawing, based on several photographs, drawn by artist William Wesley under the direction of RAS-appointed astronomers. Just as the image of the eclipse is a composite based on several records, so too will the final lithograph itself consist of images drawn on several stones, then printed on a single sheet of paper. This stone shows the corona; the prominences are on a second stone. (RGO 6/135, fol. 204r. By permission of the syndics of Cambridge University and of the director of the Royal Greenwich Observatory.)

17. A complete proof of a lithograph of the eclipse of 1871, sent to Astronomer Royal George Biddell Airy for his examination. A careful comparison of this picture and the previous shows that the prominences are now visible. Because the prominences—particularly the large one on the lower right—are far brighter than the rest of the corona, and had to be printed with a thicker ink, they had to be drawn on a second stone. A considerable amount of skill was required to draw complex images on multiple stones, and to print them accurately. (RGO 6/135, fol. 204r. By permission of the syndics of Cambridge University and of the director of the Royal Greenwich Observatory.)

18. A composite drawing of the eclipse of 1898. This lithograph was drawn by William Wesley from photographs taken by Ralph Copeland in India. Copeland decided that Wesley was the only person with the skill necessary to make the drawing. (Courtesy of the Royal Observatory, Edinburgh.)

Exposure 0·3 second; Sandell "Triple" plate.

Exposure 0·3 second; Ilford "Special Rapid" plate.

THE CORONA, 1900, May 28th.

(Photographed at Wadesborough, U.S.A., by the Rev. J. M. Bacon. Aperture
4·1 inches. Focal length 58 inches.)

19. Reproductions of two photographs taken at the eclipse of 1900. In the 1870s photographs like these were often combined into a single composite drawing for publication; now, they are reproduced without conscious alteration, with information about each photograph's exposure time, the kind of plate used, and the camera's aperture and focal length. (From E. W. Maunder, ed., *The total solar eclipse of 1900* [London: *Knowledge* Office, 1901].)

CHAPTER 4

Drawing and Photographing the Corona

Social Construction of Scientific Images

Since the 1970s historians of science have produced a number of studies arguing that, in important ways, scientific research and discovery is a process of "social construction" influenced as much by social and cultural forces as by evidence, logic, and theories. Forces normally considered "external" to science were thus given equal weight in explanations of the development and practice of science as factors traditionally considered "internal" to it.[1] Scholars drawing on the work of Thomas Kuhn, Robert Merton, and other philosophers or sociologists have expanded our view of scientists' motives, encouraged attention to the roles that instruments, technicians, and other nonscientists play in the research process, and given us a more complicated picture of the relationship between theory and experiment. More recently, among scholars raised in or converted to the constructivist tradition, the overtones of the term "social construction" have changed in yet a different way. Their studies have placed more emphasis on the "construction" side of the phrase, showing how facts are manufactured and negotiated through practices and technologies, rather than defined by interests and ideologies. This work takes for granted what was debated in the 1970s and 1980s—that social interests help shape scientific ideas—and sets out instead to show how social construction works: how people work through epistemological uncertainties, technical difficulties, and theoretical conundrums to produce knowledge that seems trustworthy. It also pays greater attention to the constraints imposed by nature on scientific practice and inquiry, though the meaning and use of the term "constraint," and the place that appeals to nature can have in constructivist histories of science, are matters of debate.[2]

Much of this recent work has been based on case studies from the history of physics, but the importance of working nature back into sociologically nuanced accounts of science is even more obvious in the field sciences. The way nature intrudes in a high-energy physics laboratory may seem complicated precisely because laboratories are designed to separate the experimental phenomenon from the outside world. Not so with fieldwork. The caprice of nature, and the impact of its animal and meteorological representatives, are more obvious when doing research with complex organisms or in the outdoors. It is also vividly clear in eclipse observation. Not only did cultural and technological constraints influence where parties went and how they lived in the field; observers also had to reckon with very stringent demands imposed by the phenomenon itself. The history of eclipse observation is the history of astronomers attempting to compress their science to fit within the bounds of an unchangeable constraint. Observing programs and instruments had to be designed to work quickly; everything that happened before an eclipse happened with this in mind, and everything that happened afterward was a consequence of it.

For visual records and representations, materials and technology provided another important set of constraints that had to be dealt with. Photographs and drawings were among the most prized possessions of an eclipse party, and it was essential that they be copied accurately. This was an exceedingly complex and difficult task, thanks to the differences between drawings, photographs, and engraving technologies, and the challenges of translating images from one medium to another. The material life of pictures, in fact, made the copying of images one of the great challenges of eclipse work. Historians of science have paid little attention to this material aspect of pictures, but not because scientists themselves were unconcerned with or unaware of the challenges of making and reproducing them. Scientific illustration has traditionally occupied an uncomfortable position between the history of science, technology, and art, and until recently the rewards of an expedition into that territory have been unclear. Further, most illustrators and engravers are anonymous or obscure figures even to specialists, and records documenting printing practices have been lost or are hard to find. But as Victorian art critic Philip Gilbert Hamerton argued, intellectual habits of mind also play a role in sanctioning a lack of interest in the making of pictures:

All of us who are supposed to be educated people have been trained in the mental habits which are derived from the study of books, and these habits, as all artists and men of science are well aware, lead students to value words and ideas more than things, and produce in their minds a sort of contempt for matter. . . . [This contempt encourages] people to think that technical matters may concern artists and still be below the region of the higher criticism, which should interest itself in the things of the mind, and not bestow attention upon the products of the laboratory, or the processes of the painting-room. . . . It even misleads our judgment by inducing us to suppose that substances are beneath the consideration of an artist, as they are outside the preoccupations of an author.[3]

Those who argue that the shift from printed to electronic text is already causing changes in the way we think and write will disagree with Hamerton's views on the unimportance of materials to the written word.[4] But his criticism shows that we cannot assume that literary "mental habits" or literary theories are sufficient for understanding the history of pictures.[5] These habits and theories are especially important in shaping our appreciation of the technological dimension of pictures. As Hamerton wrote:

In the Graphic Arts, you cannot get rid of matter. Every drawing is in a substance and on a substance. Every substance used in drawing has its own special and peculiar relations both to nature and to the human mind. . . . In literature, such a connection can scarcely be said to exist. A writer of books may use pen or pencil, and whatever quality of paper he chooses. There is even no advantage in reading the original manuscript, for the mechanical work of the printer adds clearness to the text without injuring the most delicate shades of literary expression.[6]

The same cannot be said for scientific illustrations. Images do not move easily from the sketchpad or photograph to the engraver's steel plate to the printed page. Every change of state requires a reconstruction of the image, and those moments of reconstruction are full of dangers: a photograph taken in the field is overexposed, an artist alters some critical detail while copying an original, an image dissolves in an etcher's acid bath. To write the history of scientific representation, we must enter the world of pictures and picture-making; reconstruct the constraints that scientists and their collaborators had to deal with during the creation of original pictures in the laboratory or observatory, the copying of pictures in the artist's studio, and the mass-production of illustrations for journals, atlases, and books; and examine how materials, practices, and interests were made to interact in the production of solutions to problems that constraints posed. Managing these interactions and negotiations was a particular challenge in the Victorian era, a period of upheaval and instability in astronomy, imaging technologies, and the graphic arts.[7] The solar corona was a very difficult object to observe and record, and the printing of pictures of the corona demanded attention to a variety of technical factors and raised questions about what constituted good pictures, and how the credibility of images was to be established. It drew on the skills of astronomers, engravers, and other "invisible technicians," and required the careful elaboration of lines of authority between the different groups. The very difficulty of representing the corona helps to throw the technical and epistemological issues surrounding observation, representation, and copying into bold relief.[8]

This chapter will examine the history of eclipse representation in two stages. It begins with the making and copying of images of the corona in the 1860s and 1870s. It first reviews the drawing and photographic practices used in the field by astronomers like Norman Lockyer and Warren De la Rue, as well as artists like John Brett and Henry Holiday. It then shows how pictures produced in the field were copied. It looks at astronomers' correspondence to see what choices they had in copying pictures, and how they made decisions between competing methods and techniques. I argue that in this period, astronomers seeking to produce detailed, naturalistic pictures of the solar corona coordinated a variety of skills—technical, scientific, and artistic—and brought them to bear in a process of making composite pictures of the solar corona. Composite drawings combined the best features of several photographs and drawings in pictures that were detailed, credible, and could be mass produced.

In the second half of the chapter I consider the situation in the 1880s and 1890s, when astronomers abandoned the heterogeneous mix of skills and technologies they had earlier developed in favor of an increasingly mechanical system of making and copying photographs. Professional expeditions equipped with large cameras drove amateurs and drawing out of the field. Photomechanical printing technologies like the photogravure and relief halftone offered astronomers increasingly suspicious of human judgment and observation a vision of a fully mechanical means of imaging, a closed system free of human judgment and frailty that would transmit pictures from the field to the pages of the journal. The constraints under which astronomers had worked previously seemed to have technological solutions. Large cameras permitted astronomers to do more reliable work in shorter periods of time, and new printing technologies let them escape from the errors inherent in copying images from one medium to another and the uncertainties that artistic judgment introduced during those transfers.

Representational Fieldwork in the 1860s and 1870s

The challenge of representing the corona began in the field, with the hurried attempts of observers to draw and photograph the corona.[9] Astronomers in the 1860s and 1870s relied equally on drawing and photography, and tried to develop means to improve the accuracy and detail of both. This was a period in which the division between drawing and photography, organized along fault lines between amateurs and professionals, was not very strong.[10] Photography's earliest enthusiasts included numerous artists who saw the new technology not as a substitute for human observation and representation but as an auxiliary to it. This sense that both human and mechanical observation were useful was also characteristic of early astronomical photography, whose innovators included accomplished observers and artists like John Herschel and Warren De la Rue.[11]

Three things indicate that in eclipse observation drawing was not the province of observers who could not be trusted with cameras. First, the RAS issued instructions for draftsmen just as it did for photographers. In 1870 and 1871 they were written by John Brett, a Pre-Raphaelite artist and amateur astronomer of considerable reputation, "to regulate the efforts" of RAS-sponsored draftsmen. Additional instructions for the 1871 eclipse were written by Henry Holiday, an artist and illustrator with ties to the Pre-Raphaelites.[12] Second, draftsmen were often men of scientific distinction and artistic skill, as accomplished as their photographer counterparts. Their ranks included professional artists, book illustrators, architects, military and naval officers, and engineers, a number of whom were fellows of the Royal Astronomical Society. Third, several photographers had some background or connection to the visual arts, and some artists were notable photographic experimenters. Two photographers on the 1870 expedition had family ties to the arts: Henry Roscoe's grandfather was a major writer on Medici art, while Lord Lindsay's father was an art critic.[13]

ECLIPSE DRAWING

Eclipse draftsmen sought to make accurate, detailed drawings of the corona. At first their aim was to determine whether the corona was real, and whether it was a solar phenomenon; after its existence was established, draftsmen turned to building a record of its pattern of behavior. These aims were structured by the drawing programs established by early Victorian astronomers working on nebulae. The "prime object" of astronomical drawing, Charles Piazzi Smyth wrote in 1846, was the recording of "the passage of a celestial body from one state to another." The case for nebular change could "only be established by the comparison of very exact and faithful reproductions," and Smyth had urged astronomers to develop "correctness of eye, facility of hand, and a due appreciation of the subject" and devote themselves to making "painstaking, accurate, and detailed" drawings.[14] Others seconded Smyth's calls for accurate, unvarnished visual records. John Herschel criticized some observers of the Orion nebula for "a want of sufficient care . . . in faithfully copying that which they really did see" and hence confusing rather than resolving the question of whether the nebula changed.[15]

Instructions and tools. In this spirit instructions issued to eclipse observers were quite detailed. The purpose of instructions, whether developed by artists themselves or handed down by committee, was to regulate and stabilize the work of observers, fostering a much-needed economy of effort and giving artists an anchor against excitation and undisciplined observation. The most comprehensive were developed by Pre-Raphaelite artist John Brett and painter and illustrator Henry Holiday and endorsed for RAS-sponsored parties. John Brett told readers to first note the corona's extent, whether it was concentric around the sun or moon, and whether it was diffused evenly around the sun. Next they were to scrutinize its internal structure and composition, noting any beams or "luminous patches." If observers saw beams, they were instructed to see if they were evenly distributed around the corona, note any dark rifts or spaces between them, and observe whether the corona had a definite or diffuse border. They should then continue down to the inner corona, to see if it grew brighter or dimmer near prominences. Finally, they should note whether the moon's disc was darker than the surrounding sky.[16]

Amateurs observing the eclipse of 1878 were issued similar instructions. They were told to note the color of the corona, then record if its rays were straight or curved, whether there were changes in the brightness of the rays and whether changes were sudden or gradual, whether the corona was "a mass of diffused light" or made up entirely of separate streamers, and finally to note how far the rays extended.[17] Other artists favored a more opportunistic approach, first locating the most interesting features of the corona with a few seconds' observing, then rapidly sketching the entire corona, then returning to the eyepiece, this time to concentrate on those most important features. This required steely nerves and an ability to think fast but had the great advantage of flexibility. Others drew everything in a single quadrant of the corona, from prominences to outer wings. Finally, a few targeted the outer corona, and blocked out the inner corona with occulting discs mounted on a pole.

Most draftsmen worked at desks or easels beside modest-sized telescopes. Like all eclipse fieldworkers, they spent the hours before totality making final preparations, checking instruments, laying out pencils, pen, or charcoal, and drawing black discs and polar coordinates on their cards and paper. Those concentrating on the outer corona mounted an occulting disc on a pole to block out the inner corona. A few minutes before totality many covered their eyes with dark cloth to make them more sensitive. Once totality began, they worked as quickly as possible, observing the corona for a few seconds, then rapidly sketching what they saw, then returning to the eyepiece. Many planned first to make preliminary drawings of the entire corona, then additional sketches of the more interesting features; others observed and drew only a single quadrant.[18] Most worked in a self-consciously unadorned style. Piazzi Smyth had criticized the "misplaced attempt to produce a splendid picture" full of "high finishing" that generated many beautiful but scientifically useless pictures, and artistic refinement was distrusted for its willingness to alter natural appearance.[19]

Excitement and accuracy. However, even the best-prepared observers were sometimes so nervous during totality that their plans could not be carried out. Architect R. F. Chisholm planned to sketch the prominences, but "unfortunately my hand shook so with . . . excitement that I could do little more than indicate their position." Henry Holiday wrote, "I could with difficulty control myself so as to be fit for making a decent observation," and immediately after totality he retreated to his room, "where I first plunged my head into water, for I was in a fever of excitement."[20]

This was a problem for all observers, regardless of what tools they used. John Herschel, who set out to observe the spectrum of a prominence in 1868, presented a vivid portrait of the excitement of an eclipse and the difficulties it produced. Before totality he was gripped by a "fever of philosophical impatience" that made last-minute adjustments difficult. Then, seconds after the eclipse, a cloud moved over the sun and its image was lost. Herschel had to put down a "frenzied temptation to turn screws and look somewhere else," and after a few seconds the cloud passed. He found a large prominence and immediately saw that the spectrum consisted of bright emission lines. "I think I was a little excited" by the discovery, he continued, "for I shouted to my Recorder, 'Red! Green! Yellow!' quite conscious of the fact that I meant orange and blue. I lost no time in applying myself to measurement," but by third contact he had managed to measure only one of three bright lines. "I have no idea how those five minutes [of totality] passed so quickly," he told William Huggins afterward.[21] After reading Herschel's report George Airy felt that he had been "excited to a degree unfit for good observation."[22]

Finishing drawings. No one tried to make a finished drawing during an eclipse. During totality observers sketched the broad outlines of the corona and some of its more important features, and made more detailed drawings after totality that included details they observed but did not have time to record. From the moment of their creation, drawings made during totality had a status different from "finished" drawings. As Francis Galton explained, the corona "was too manifold in its details and too beautiful in its proportions . . . to do justice to it in the short time the spectacle lasted." Eclipses were so striking that observers saw and remembered a tremendous amount about them. Warren De la Rue claimed that while "not more than 20 seconds were devoted to observations with the unassisted eye" at the eclipse of 1860, "the phenomena remain strongly impressed on my memory," and even a year later he could remember it "as vividly as if it had but just occurred." Despite his nervousness during totality, R. F. Chisholm saw well enough to make a detailed drawing of the corona after the eclipse, explaining that "anyone accustomed to sketch from nature could with ease have retained the scene by memory." Members of a party organized by the Indian Trigonometric Survey in 1868 drew sketches of the prominences during totality, then "aided by our rough original sketches, and our memory, each made another diagram" directly after the eclipse. George Darwin made a quick sketch of the entire corona during totality, a "partially finished sketch" just after the eclipse, and a color drawing a few hours later. This practice was even recognized in some drawing instructions. James Tennant advised Indian amateurs how to observe the eclipse "so as to enable drawings to be made and descriptions to be written immediately afterwards."[23]

The first attempt to photograph an eclipse was made in 1851, using a daguerreotype camera. Warren De la Rue was the first British astronomer to photograph the eclipse, using the Kew photoheliography and collodion plates in 1860. The number of observers photographing the corona and eclipse spectrum steadily increased through the century, as they abandoned the wet collodion in favor of dry collodion and then gelatin plates.

Most expedition photographers in the 1860s and 1870s used glass plates coated with wet or dry collodion and photosensitive silver iodide. Wet plates were prepared by cleaning a glass plate, smoothly and evenly pouring collodion (a viscous nitrocellulose dissolved in alcohol and ether) mixed with potassium iodide or bromide onto the plate, then immersing the plate in silver nitrate solution for several minutes to produce a coat of photosensitive silver iodide. Great care had to be taken in preparing wet plates: the glass had to be very clean, and the solution free of smudges, bubbles, or streaks. Further, collodion was highly unstable if incorrectly prepared: ether is flammable, and nitrocellulose is the active ingredient in smokeless gunpowder.[24] Wet plates had to be prepared minutes before second contact, and dry plates backed with a light-absorbing film to prevent "halation," reflection of light within the plate.

Regardless of the type of plate used, under ideal conditions each camera in an eclipse party would be manned by several people. In reality, observers often had to "attend two or three photographic telescopes" alone, and complaints about shortages of "intelligent assistants" were common.[25] Ideally, however, one assistant would fix the prepared plates in a holder and hand them to the photographer. The photographer would insert the holder into the camera, make any necessary adjustments to the telescope or spectroscope, and determine the exposure times. Another assistant would stand at the head of the camera or telescope, covering and uncovering the opening to expose the photographic plate. A third assistant would take the exposed plates from the photographer and carry them to a darkroom. Wet plates had to be developed on the spot, since undeveloped plates would be ruined if the collodion dried. They would be developed in a bath of ferrous sulfate, which washed away the undeveloped silver iodide; the developed plate was fixed with sodium thiosulfate or potassium cyanide, washed in distilled water, and dried over a candle or burner, then covered with a coat of lacquer.[26]

Reproducing Photographs and Drawings
in the 1860s and 1870s

The work of representing the corona did not end with third contact. As Smyth indicated in his 1846 article on astronomical drawing, astronomers with good pictures now had to confront the "hardly inferior" challenge "of getting them engraved."[27] Good reproductions had to be realistic, detailed, and made in a fashion that was trustworthy. Paper positives could be produced in small numbers, but they were expensive and lost contrast and detail. Bad copies could lead to bad conclusions and embarrassing mistakes. Richard Proctor attacked Norman Lockyer's claim that photographs showed the nonsolar origins of the radial structure of the corona with the withering: "[T]he radial structure is not lost in the photographs. It is lost in the positives, but in the original negatives it is perfectly manifest," as if this were an elementary mistake.[28] Privately made copies were suspect because their means of production were unknown (or unsupervised), and a viewer could never be sure that one had not been altered or retouched.[29] Producing pictures for one's colleagues required deciding how field-produced drawings and photographs should be copied, choosing between different printing media, finding skilled engravers and printers, negotiating access to originals with astronomers and expedition sponsors, and making the process subject to public testimonial and approval. But the process of transferring images from the field to the printed page was not looked upon simply as a chance for error and artistic ambition to infect scientifically valuable pictures. Rather, some Victorian astronomers saw an opportunity for improving the quality of images, and producing pictures that combined all the details and features of originals and none of the problems.

PRINTING TECHNOLOGY

AND REPRODUCTIVE AESTHETICS

Victorian astronomers worked in a period in which printing technologies flourished, debates over the merits of drawing versus photography were rampant, and the standards by which originals and reproductions were judged were in flux.[30] Photography was changing the standards by which illustrators made pictures and the public judged them, but until the 1890s and the development of relief halftone, copies of photographs could not be mass-produced. The history of art reproduction illustrates these problems and provides an especially useful comparison to scientific publishing.[31] Like astronomical objects, paintings and sculptures are complicated subjects, and good reproductions were made for informed and demanding audiences. Knowledgeable collectors did not expect reproductions to be "exact" copies of art works, judging them instead on their ability to capture the spirit of the original. Engravers' professional identities were based on the idea that they were not merely copyists and clever mechanics, but translators with unique and specialized skills.[32] Both producers and consumers of art reproductions looked upon them as reinterpretations that could be appreciated as minor works of art. The flourishing of photography, and the emergence of a high-end market in collotype reproductions, began to undermine this system in the 1870s. By the 1880s it was in turmoil. Some illustrators continued to print idealized engravings of sculptures and works of architecture. Others broke with the interpretive and idealist schools and set out to develop an "art of illusions" in which the tone and texture of oil paintings were reproduced in engravings. Still others turned to engraving facsimiles of photographs. This confusion of standards and aims was matched by a confusion of illustration and printing processes, as books were filled with photographs of engravings, engravings of photographs, and so forth.[33]

CHOOSING A PRINTING METHOD

Astronomers had first to decide what sort of visual evidence they wanted to lay before their readers. Exact reproduction was an effort to place before readers the same evidence that astronomers worked from. When De la Rue prepared his report of the eclipse of 1860, George Airy insisted on the importance of reproducing originals without alteration. As Holly Rothermel has argued, "Airy saw photography as a tool to replace and standardize observers, not as a method of discovery or invention," and his views about reproduction were just as strict.[34] "Exact facsimiles of the two eclipse photographs" De la Rue took were, he argued, "infinitely more important" than composites or retouched plates.[35] Even a photograph that was "practically useless unless touched up must not be omitted," for it,

and not the touched-up photograph, contains the evidence about the case. The interpretation, on which parts are stopped out, *may be* fallible (I do not believe that it *is*), but the whole question about the prominences is strongly debated, and you must proceed with exactly the same caution as in a disputed case in a court of law.[36]

De la Rue made two facsimile engravings of his photographs, but it was very difficult work. Just copying the photographs took months: it was "tolerably easy" to make copies "which show the results fair well" when examined by eye, but copies for reproduction had to be stronger and more carefully made.[37] A decade later James Tennant would make a similar argument when his own coronal photographs were being reproduced. "*We* may be satisfied" with the appearance of a retouched photograph or composite drawing, he wrote, "but we are bound to provide if possible our evidence [to our readers]. . . . Each photograph should be reproduced in facsimile as far as possible and the artist should have no power of impressing himself [in the picture]."[38]

By the time Tennant wrote it was much easier to make exact copies of original photographs. London had become home to a number of collotype printers, working under various trade names such as heliotype and Albertype, who could have reproduced eclipse photographs. De la Rue himself was a pioneer in applying collotype to astronomical illustration.[39] Collotypes were made on glass or thick metal plates coated with a light-sensitive mixture of gelatin, potassium bichromate, and alum. A photographic negative attached to the plate and illuminated from behind would be recorded on the plate as a series of hard and soft areas, the hard areas corresponding to the lights in the negative. Pictures were usually printed directly from the gelatin. The plate was first wet with a sponge, and the water was absorbed in the least-exposed areas. A greasy ink was then applied to the plate, which adhered to the gelatin in proportion to its degree of exposure and water absorption.[40] A well-made collotype print was a masterpiece, with considerable detail, good contrasts, and "soft and continuous [tones] showing no grain to the eye." The drawbacks of the process were that the blacks were not very deep, the edges of the picture were sometimes slightly distorted, the process was easily affected by humidity, and print runs were short. Further, unless the negatives were retouched, stains and flaws in the original plate would be printed along with the corona.[41]

But even as collotype was attracting attention in photographic circles, a new attitude toward eclipse pictures was emerging. This contrasting position held that field-produced pictures were incomplete visual statements and that more would be gained from publishing images that were the product of both fieldwork *and* analysis. It was widely accepted that photographs fell far behind paintings and engravings in their ability to present nature in all its variety. Art critic Philip Hamerton wrote: "It is one of the peculiar misfortunes of the photograph that it is not capable of giving two truths at once . . . not having any method of compensation like that which every painter finds out for himself."[42] This was true of eclipse photographs, for the tremendous differences in brightness between the inner and outer corona meant that one part of the eclipse could be recorded only at the expense of another.[43] Worse yet, by varying the background lighting or developing times during the copying process, new details could be discovered that could not be captured in the same reproduction.[44] For example, James Tennant made a drawing of a "horned protuberance" that showed it "not perhaps as seen at any one time, but as it exists in the Photographs, and is brought out by varying the illumination."[45]

A photograph could not give "two truths at once," and a direct reproduction could not even copy everything visible in original photographs. Better that the engraver's fees be put to producing an image of the eclipse as it really was, not as it appeared on individual photographs. Such a picture might be distilled from a close analysis of a number of photographs, yielding an image that showed the entire eclipse—prominences, inner corona, outer corona—at once. Some photographers went into the field with this procedure in mind, constructing practices that would let them get around the constraints of their technology. Manchester photographer Alfred Brothers planned to take a series of photographs of the eclipse of 1870 "from which a picture of the corona could be, so to speak, built up."[46] Another team reported from the field that they had secured nine photographs from which "a drawing can be made that will show the whole structure from the limb to the farthest extensions of the corona."[47] Astronomers could scrutinize photographs and see how different features were represented in different photographs, but few had the artistic skill to turn these observations into reproducible pictures. For that they had to enlist the aid of engravers and printers.

The introduction of engravers was only one complication in the process of "building up" pictures of the corona. Astronomers set high standards for these pictures in terms of accuracy and naturalistic appearance. Simple woodcuts could suffice for newspapers and popular books, but to be scientifically useful engravings could not have visible lines and crosshatchings. The fact that the corona is a tough object to draw realistically—in addition to its range of brightness, it is at once nebulous and diffuse, with a complex internal structure of rifts and arches that had to be represented—and was unfamiliar to most artists did not improve matters.[48]

BOUNDARIES OF AUTHORITY

Further, astronomers and engravers each had their own ideas about what constituted a good reproduction, and who had the competence to define good pictures. Engravers were skilled workers who saw themselves as "translators" and "interpreters" and who were accustomed to a measure of independence in their work. Attempts to dictate the appearance of engravings, or even to monitor their work too carefully, represented a challenge to their competence and skill. Astronomers, on the other hand, took for granted that they knew what astronomical pictures should look like, and worried about losing control of the process to hired hands. Photographs could not be naturalistic pictures, or completely trustworthy ones; the opportunity existed to create such pictures when transferring them from field sketchbooks and glass plates to journal pages. The challenge was to create a system that could use existing printing technology, draw on the skills of both astronomers and engravers, and establish divisions of labor and authority that would be acceptable to both parties. Engravers had to be monitored and corrected, especially when working at one or more removes from original pictures. As Charles Piazzi Smyth wrote: "Errors are always copied, and magnified as they go . . . [so that] after a few removes, the alleged portrait of nature is only a caricature of the idiosyncrasies of the first artist."[49] But the engraver's skills needed to be enlisted if copying were to be successful. The problem was that the line between improvement and introduction of the personal equation into the final product was always thin, and agreement over what actions fell on either side of that line, though they would be open to challenge at some level, had to be worked out. The credibility of images would depend ultimately on the credibility of the technological and social system that produced them.

A CASE STUDY: THE RAS MEMOIRS ECLIPSE VOLUME

A detailed record of how eclipse pictures were made is available for the most comprehensive records of eclipse observations published in the 1870s. Volume 41 of the Royal Astronomical Society's *Memoirs* was devoted to eclipse accounts and observations since 1715. Its history reveals how in a period in which nothing in the production process could be taken for granted, accounts and records were evaluated, photographs and drawings were examined and treated, and recording technologies were weighed and choices made between them.

The six-hundred-page-long volume was the most lavishly illustrated eclipse volume ever published, containing hundreds of woodcuts and two dozen plates. When George Airy began work on the volume in 1871, it was a fairly modest collection of accounts and drawings of observations from the eclipses of 1860, 1869, and 1870. When duties at the Greenwich Observatory kept him from continuing the project, however, Arthur Cowper Ranyard volunteered to take it over, and it was under his editorship that the volume expanded to include observations from as early as 1715.[50] Ranyard, a Cambridge graduate and lawyer by profession, was no stranger to RAS affairs. He was already a member of the council and had served as secretary of the committee that organized the 1870 eclipse expedition.[51]

Ranyard spent much of 1871 compiling and editing accounts of observations, and corresponding with observers of recent eclipses about technical details and aspects of their reports.[52] As early as January 1871 John Brett had asked him how the drawings from the 1870 eclipse were to be engraved.[53] He turned to the subject in early 1872. Ranyard and Airy agreed that woodcuts "of the more remarkable published drawings of the corona"—indicating that the two assumed originals would be difficult to get—would be included in the text. These would serve as accessories to written descriptions. A second series of plates was made from photographs of recent eclipses and would be included at the end of the volume, unencumbered by any text.[54] Ranyard estimated that small woodcuts would cost about thirty shillings each, and that he could easily spend £150 to £200 on illustrating the volume.[55] Airy wanted the report to be exhaustive. "Even the worst [photographs] of all are not to be suppressed, but may be verbally described," he told Ranyard; simple drawings were to be made of decent photographs, and no expense spared for the best images.[56] Images of the same eclipse were made to the same scale to make comparison easier. The ideal size included a lunar disc one and a half inches in diameter. The image would be large enough to include all the important details, but small enough to "permit . . . two facsimiles being on one plate."[57]

Making composite drawings. In February 1872 Ranyard hired William Henry Wesley to execute the woodcuts and make the drawings from which the large plates would be engraved. Wesley was recommended by William Huggins, who noted that he had "done many woodcuts in the recent numbers of the *Philosophical Transactions*."[58] Though he has been forgotten even by historians of Victorian science, in

1870 the thirty-year-old Wesley was one of the best known scientific engravers in London.[59] His family was part of the community of booksellers, chemists, mineralogists, printers, and other merchants and tradesmen who supplied London scientists, museums, and societies.[60] William Henry was apprenticed to an engraver at fourteen, and turned to scientific illustration after meeting T. H. Huxley in 1862. Through Huxley he was introduced to and eventually worked for Joseph Barnard Davis, Sir John Lubbock, St. George Mivart, and Herbert Spencer.[61] He also began working for Robert Owen, and would illustrate Owen's papers until the naturalist's death.[62] By the end of the 1860s, in short, Wesley was illustrating both sides of the bitter debate over evolution of species.[63] It took Ranyard time to convince Wesley to accept the assignment: he had recently married, opened a studio, and wanted to leave scientific illustration for painting, but eventually Ranyard prevailed. In 1875 Wesley became RAS assistant secretary, a position he held until his death in 1922.[64]

Wesley engraved sixteen plates for the *Memoirs* volume. Seven were based on photographs or drawings of recent eclipses. All of the plates credit Wesley and the printer (most were printed by Malby and Sons, who also printed many plates for the *Philosophical Transactions*), and describe their sources. Four of the plates of earlier eclipses were drawn from engravings of original photographs, or from glass or paper positives. They show far less detail than three plates from 1870 and 1871, which were drawn by Wesley and Ranyard from the original negatives.[65] These originals were kept at the Greenwich Observatory or were loaned out to De la Rue, Lord Lindsay, and William Huggins. They were hard to recover, for they were in demand by exhibitors and researchers.[66] Plates taken by astronomers who had observed eclipses on their own were still in their owners' possession; those taken on officially supported expeditions were the property of their sponsors, and were controlled by Airy.[67]

Ranyard and Wesley first worked on Alfred Brothers's 1870 photograph and Lord Lindsay's series of five photographs taken at Baikul, India, in 1871. They began by compiling inventories of all the features visible in all the photographs taken at a particular site, describing the details as they appeared in the photographs, accompanied sometimes with drawings by Wesley.[68] Ranyard then looked for Tennant's 1871 photographs, but found that De la Rue had just taken them.[69] De la Rue refused to turn them over, and Ranyard instigated a confrontation at an RAS council meeting. Airy interceded, dressing down Ranyard and apologizing to De la Rue. After a couple more days of Airy's stroking, De la Rue grudgingly handed over the plates.[70]

Once he had them, however, he did not turn them over to Wesley; Ranyard insisted that the artist work in his apartment, for he refused to let the irreplaceable originals leave his possession.[71] Airy had warned Ranyard about the consequences of risking damage to a plate by "letting an original go into the hands of a mere tradesman"; and besides, artisans were only occasionally allowed to work with originals, and only then under close supervision.[72] De la Rue's 1868 engraver made his basic drawings from paper positives, and had to travel to Greenwich to make final corrections from the originals, working at "a small table near a window looking toward the North" under the eye of a Greenwich staff member.[73]

The drawings Wesley made were not copies of individual photographs but composites of photographs. This was a critical point in the reproduction process, one in which the skills of astronomers and artists all had to be delicately balanced to yield improvement rather than interpretation.[74] The process began with Ranyard and Wesley compiling their inventory of details of photographs taken at a given site. Since the pictures were so small—as Ranyard put it, "the whole extension of the corona could be covered by a sixpence"—this was a very difficult and time-consuming process.[75] (The small size of the negatives probably made them impossible to reproduce by collotype.) Further, because of the pictures' small size, even a speck of dust or grain of sand on the plate could have a substantial impact on the quality of the image. As a result, Ranyard explained,

it would be impossible from the examination of a single negative to determine whether any small marking has its origin in some almost microscopic impurity on the collodion, or whether it represents a vast mass of many million cubic miles in the corona; it is only by a careful comparison of the different negatives that such photographic flaws could be properly eliminated.[76]

A detail had to be visible on at least three of the five plates to be certified real. When printed, the plates carried captions assuring readers that "no detail is included that could not be seen in at least three photographs."[77]

Since they were working with negatives, they concentrated their attention on "the dark or partially opaque details of the photographs which correspond to the luminous details of the corona"; in contrast, bright "spots, lines, or patches" in the photographs, which corresponded to dark features, were "regarded as photographic defects." This made them overlook certain interesting features in the corona that appeared as lights in the negatives and were thus regarded at first glance as artifacts, they later realized.[78] Ranyard probably supplemented these studies with an examination of drawings and written descriptions of the corona, which he was collecting at the time. In the composite drawing stage, photographs, drawings, and direct observation existed in a symbiotic relationship. Warren De la Rue had made eye observations at the eclipse of 1860 "in order that I might be in a position to interpret from my own sketches and recollections the results of the photographs," and plates of the eclipse were "compounded of my own drawings and the photographs."[79] It was time-consuming and demanding work, pursued by Ranyard in hours stolen from his legal practice. The 1871 photographs alone took a full year to catalog.[80]

Wesley began his drawings by making separate outlines of the corona from each of the best photographs. The outlines were superimposed, the points of disagreement between them examined, and a final composite outline was drawn. He then filled in the details and prominences, using the inventory and its drawings. A lithograph of Lord Lindsay's 1871 photographs was finished in June, and Ranyard invited Airy to study them in Lindsay's apartment.[81] Ranyard and other astronomers would have the chance to compare the composites to the originals, but not until Wesley invited them to do so. This gave the artist a measure of freedom but allowed the astronomers back in at a mutually satisfactory moment to correct and certify his work. This was a standard division of authority and labor. Drawings of James Tennant's 1868 photographs were begun by a draftsman in the Surveyor's Office working alone, and finished after Tennant drew the artist's "attention to defects of representation, or over-definition."[82] Sometimes it turned into an exercise in multiple witnessing. Corrections to engravings of photographs of the 1868 eclipse were made "in the presence of" Warren De la Rue, Astronomer Royal George Airy, Admiral R. H. Manners, and RAS Secretary E. J. Stone. The work was considered finished when all four "expressed themselves satisfied with the results." This system also served to ensure the artist's objectivity: Tennant declared that James's drawings "may be considered as the work of a conscientious copyist devoid of all theories."[83] Composite pictures were intended to combine artistic and scientific judgment in the production of an image that had all the virtues and none of the flaws of original photographs or drawings. A skilled artist, working from good originals and with astronomers, could produce a clearer and more truthful picture than any hurried draftsman or photographer working in the field. Astronomers were not willing to sacrifice beauty and accuracy for mechanical autonomy, at least not yet. Neither, I think, did they believe it possible or desirable to cut out engravers from the reproduction process. A talented man like Wesley was a greater virtue than he was a threat, an essential ally in the effort to produce superior images of the sun.

Printing the composites. Once the composites were finished, Ranyard turned to the problem of printing the plates. His summary of the advantages and disadvantages of various methods of producing eclipse pictures was presented to Airy in September. All printing technologies fall into three classes: relief, intaglio, or planographic. In relief prints, the raised rather than depressed areas are covered with ink and brought into contact with paper to produce the image. Intaglio prints hold ink in furrows cut into a plate, and paper is pressed onto the plate, absorbing the ink. Planographic prints use chemical rather than mechanical processes to hold ink. Intaglio and planographic prints were never used by magazines and books that wanted illustrations on the same pages as printed text, since text was always printed in relief. Separate plates were more expensive, but they cut production time by allowing compositor and engraver to work independently, and allowed plates to be printed on higher-quality paper.

The three most promising methods, Ranyard thought, were steel engraving, lithography, and heliotype. All three were cheaper than photographs and could sustain a wide range of tonal contrasts. Metal relief etchings were made by drawing on a plate with an acid-resistant ink and then bathing the plate in acid to etch down the blank areas. Engravings would record small, faint details in the corona that other methods might lose, and the "delicacy of the very faint lights" would do justice to "the faint edges of the corona." However, the "minute lines" of the engraving became visible upon close scrutiny, and the impressions were made on damp paper that shrank, altering the scale of the drawing. Steel engravings would also be three times as expensive as lithographs or heliotypes.[84]

Ranyard then considered lithography. Lithographs were made by drawing with a grease pencil or chalk on a rough drawing stone. The lithographer worked at a special desk equipped with a turntable (making the heavy stones easier to move) and board over the stone on which the artist could rest his hand. If copying, the lithographer would place the original under a mirror and draw the reversed image. Once the artist was finished, the stones were cleaned with a fine brush and coated with water, which soaked into the exposed areas. Lithographic ink, applied with a roller, was greasy, and would be repelled by the water and adhere to the drawn areas; the image would then be transferred by pressing paper against the stone. Great care had to be taken in preparing and drawing on the stone, since fingerprints, sweat, and skin would ruin the appearance of finished prints.[85] Midcentury lithographers were stylistic chameleons, able to draw pictures that resembled engravings, mezzotints, chalk, and oil paintings.[86] Ranyard noted that lithographs would be cheap and even better at capturing tonal contrasts than steel engravings, could be printed on dry paper, and did not have the distracting cross-hatching. However, the corona and prominences would have to be drawn on separate stones (since the absolute white of the prominences required heavier ink), "and unless this is very carefully done there is apt to be a slight displacement."

Heliotype was a variety of collotype. Since it was a photomechanical process, Ranyard told Airy, "the work of the artist who measures the original negative is exactly copied without being 'translated' or similarly altered as it often is by the workman who actually engraves the steel plate." However, the blacks were "not by any means absolute," and the edges of the picture would be slightly distorted.[87]

Ranyard must have discussed the problem with fellow RAS members, for other astronomers were expressing their opinions at the same time. James Tennant warned that lithographic stones "will not give 1000 copies . . . without sensible deterioration," and argued in favor of steel engraving.[88] De la Rue thought that mezzotint was "particularly well adapted to render truthfully and with peculiar softness the graduated lights of the corona," but it faded over time and could not sustain small details. He thought that steel engravings would present "minute details of form," but could not communicate subtleties in shade and contrast.[89] Airy thought that "the cross-hatching is conspicuously seen" in steel engravings, and that lithographs offered the best compromise.[90]

Working with the engravers. Ranyard settled on making both lithographs and steel engravings of the three best plates. Daniel J. Pound and James Scott were hired to make and print the engravings.[91] Little is known about Scott, but Pound had made his reputation as a portrait engraver and was familiar to De la Rue. Between 1858 and 1863 he engraved hundreds of portraits of the Royal Family, parliamentarians, divines, and scientists for the *Illustrated News of the World*. By 1861 more than 4 million copies of his engravings had been sold. From this work Pound had become familiar with the techniques of engraving from photographs.[92]

Pound laid down his conditions before taking up the work. He would base his preliminary engraving on drawings made by William Wesley, as long as those drawings were based on the original photographs and drawn to the size desired for publication. They would be "something to work from" in the early stages, but once he was ready to add the small details into his engraving, he would need access to the original negatives. At that point, Pound would move to De la Rue's house, where the negatives were now kept, allowing him to check the engraving against the originals while keeping them in the astronomer's care. "When the plates are sufficiently advanced," Pound concluded, Ranyard and others "could compare and correct the engravings from the original photographs." Each plate would cost about £25 to engrave, and £1 to produce a hundred copies on "paper of the best quality."[93] The artist also offered to proof the plates for £4, "an expensive and often-repeated operation," but absolutely necessary.[94] The total cost of producing one thousand copies of a steel-engraved eclipse, Ranyard estimated, would be £46: £7 for Wesley's lithograph, £29 for a proofed engraving, and £10 for one thousand copies.[95] Ranyard was eager to get started. As soon as the technical issues were settled and money secured, he told Airy, "I shall . . . push on the work merrily."[96]

Ranyard spent the next year overseeing the engraving and correction of the plates. Most of his time was spent getting drawings from Wesley, checking and correcting Pound's and Scott's engravings, and responding to Airy's demands that certain observations be given special attention. Production difficulties, financial complications, and countless small problems at the printer's and bindery absorbed another four years.[97] An 1874 paper in which Ranyard described some interesting coronal details that he and Wesley discovered in the course of compiling their inventory whetted his audience's appetite. "The amount of detail which Mr. Ranyard and Mr. Wesley together have discovered in the corona is most astonishing," an astronomer at the Royal Naval College remarked to Airy.[98]

When it was finally published in 1879, the *Memoirs* volume brought Ranyard considerable praise. American amateur Henry Draper wrote that "the amount of labor you have put upon it is perfectly tremendous." Samuel Langley told Ranyard, "I do not know of anything more valuable than the work you have carried through." Harvard College Observatory director Edward Pickering and Astronomer Royal for Ireland Sir Robert Ball both predicted that it would become a standard reference for astronomers.[99] Richard Proctor used it as his principal source in an article on the relationship between sunspots and coronal appearance.[100] In short, Ranyard succeeded in controlling the problems in copying and circulating field-produced images of the corona, and produced one of the few volumes that could serve as a tool for scientific research.

The history of the *Memoirs* volume highlights the contributions made by different people with different skills to the production of astronomical images. It shows how astronomers thought about photographs and drawings, what they thought constituted a worthwhile picture, and how they approached the problems of making and copying those pictures. Opportunities to create different kinds of pictures using different technologies were noted but rejected. It also created new skills in evaluating and working with astronomical pictures. Arthur Ranyard went on to edit the journal *Knowledge*, which under his direction drew praise for the quality of its illustrations.[101]

More notable was the authority it gave to William Wesley as a judge and copier of coronal pictures. Wesley's official position in the RAS hierarchy was rather modest. As assistant secretary he was in charge of the library and managed the society's correspondence, announcements, and daily affairs. Even so, senior astronomers sought out his advice on eclipse matters and asked him to write comments on their photographs. William Huggins turned to Wesley for an opinion of whether his attempts to photograph the corona outside an eclipse were successful.[102] Wesley also continued to be a popular illustrator. A number of woodcuts and lithographs—it is impossible to know how many, since some were unsigned—were published in the *Monthly Notices*, and he occasionally published in other journals.[103] Wesley also lithographed maps of the Milky Way and moon.[104] Sidney Waters, in an article accompanied by one of Wesley's last plates, thanked him "for the great amount of trouble he has taken to secure a satisfactory reproduction of the charts, and in particular for very skillfully lithographing the Milky Way upon the reduced scale."[105] Oxford professor Herbert Hall Turner deferred to the assistant secretary when working on the *Monthly Notices'* illustrations: "[If] I am overruling your own judgment," he told Wesley when working on drawings of Mars, "this is what I don't want. You know more of planetary drawing work than I. Please let me know if you would rather proceed as you suggested."[106] But eclipse drawing remained Wesley's specialty, and he was its undisputed master. Ralph Copeland described him in 1899 as the "one expert capable of making ... drawings [of eclipses] in the most satisfactory manner," and waited several years for Wesley to make a composite of his plates.[107] Arthur Common simply said that, in eclipse drawing, "[y]ou are absolutely safe in Wesley's hands. . . . [He] knows best."[108]

The system astronomers developed in the 1860s and 1870s for producing images of the corona, and then reproducing those images for publication, represents an attempt to deal with a variety of constraints. The eclipse imposed constraints of time and geography, and the instructions for eclipse drawing, and the techniques developed for eclipse photography, were designed to make the most out of the very limited time fieldworkers had to observe the corona. Astronomers had to choose carefully what they were going to do, and just as important, what they weren't. Even the use of drawing and photography was justified on the grounds that each had its own special strengths, and could succeed where the other fell short. Some photographers responded to the constraints imposed by their medium's technical limitations with the practice of composite photography, in which they varied exposure times to produce a series of plates from which a composite image could be later extracted. Not all the work had to be done during totality, either. Draftsmen could finish their sketches in the hours after an eclipse, when the event was still fresh in their minds, while photographers could work with artists to produce composite drawings of photographs months or years later, even when methods were available that permitted more direct mechanical reproduction of original photographs.

The Astronomical Image in the Age of Photomechanical Reproduction

Eclipse observers in the 1880s and 1890s faced many of the same constraints in the field as their predecessors, but they dealt with them in very different ways. In fact, the period witnessed the erosion of both the status of eclipse drawing and the abandonment of the system of composite drawing that brought together astronomers, engravers, and multiple copies of drawings and photographs, in favor of a much more centralized and mechanized system. Fieldwork became more high-tech and photography-intensive. Expeditions went into the field with custom-made instruments designed specifically for eclipse use that produced larger and more detailed photographs of the corona than their predecessors, while drawing was pushed to the margins of eclipse fieldwork. Composites declined because the development of photogravure and halftone printing made it possible for pictures to be printed directly from photographs, and offered astronomers the promise of perfect mechanical reproduction uncomplicated by human intervention either in the field or in the studio.

ASTROPHYSICS BECOMES BIG SCIENCE

The last decades of the nineteenth century saw major changes in the character of astronomy in Europe and North America. Astrophysics became institutionalized as a discipline distinct from astronomy. Large telescopes, improved mountings and drives, and dry collodion and gelatin bromide plates improved the sensitivity and range of astrophotography. The longer training, increased financial and institutional resources, and specialized skills required to do astronomical research opened a gap between amateur and professional astronomers. Older astronomers like Huggins who came out of the amateur tradition continued working, and a few younger amateurs made contributions to the field, but more and more they were in the minority: each new generation of astronomers was more highly educated, more academic, and more specialized than the last.[109] There were also more basic changes in the ways astronomers made observations, and in the ways they thought about photographic and other mechanical versus nonmechanical records.

EVOLUTION OF INCOMMENSURABILITY

The changes did not take place entirely at the level of careers and institutions. The relationship between photography and drawing, which had previously been treated as rough equals, changed dramatically in the 1880s. This was part of a broader eclipse of human observation by mechanized technologies of observation and imaging in the late Victorian era described by Lorraine Daston and Peter Galison. They argue that a central characteristic of late-nineteenth-century science was a search for mechanical substitutes to human observation, and a growing distrust of forms of visual representation dependent on human judgment.[110] Two things in particular drove a wedge between photography and drawing.

First, astronomers produced a series of dramatic and highly publicized photographs of star clusters and nebulae far superior to anything produced before. These photographs were immediately seen as superior to visual records produced by painstaking and careful researchers; they settled debates over the existence of hard-to-see deep-space objects; and, for the first time, they could be used in research requiring precise measurement. Photographs of the Orion nebula taken by Andrew Common and Isaac Roberts in the early 1880s showed faint details never seen by the naked eye and not recorded in even the best drawings.[111] Photographs of the Pleiades nebula revealed even more dramatic differences between visual and photographic studies. Wolf's chart of the area, the result of more than three years' careful labor, was made obsolete with a single two-hour exposure. Not only did photographs seem to settle debates about the magnitude and position of stars that some accused Wolf of mismeasuring, they also showed three times as many stars as Wolf observed. The photographs also settled the question of the existence of a nebula near Maia which had vexed astronomers since 1859, and they revealed another nebula around Merope. In a few years, the entire Pleiades seemed to be woven together by an immense nebular substructure. These discoveries were made all the more dramatic by the fact that they could be reproduced by numerous photographers but could not be directly observed, even with the most powerful telescopes. Many of the faintest stars "can be seen with no telescope, for the eye is dazzled by the light of the bright stars" nearby. Similarly, the nebula was too faint and diffuse to be seen by the human eye, but the photographic plate could record it with ease.[112] These photographs were all the more dramatic because the Orion nebula and the Pleiades had been the subjects of literally centuries of scrutiny, and were so familiar to every astronomer.

Second, astronomers and commentators began to talk about photographs and drawings as being fundamentally different and incompatible. Previously, most astronomers described the two media as complementary, each possessing strengths that the other lacked. The sense that both were useful was reinforced by the fact that many photographic experimenters were also skilled artists. Now, however, cameras were so much more sensitive than the human eye that comparisons between the two began to seem pointless. Unlike a drawing, which was obviously the imperfect product of fallible human labor, the glass plate was both infinitely patient and completely passive—the best of all worlds for astronomers. "Stars should henceforth register themselves," the *Edinburgh Review* declared, a phrase that suggested stellar compliance and photographic neutrality. By 1891, photographic experts like Isaac Roberts saw stellar and nebular photography as the yardstick by which other kinds of astronomical observations were to be measured.[113] While photography was improving dramatically almost year by year, drawing was not. Drawings, no matter how carefully they had been made, could no longer be trusted as accurate historical records of the state of the sky. Hence, in 1888 one magazine wrote that Common's 1883 photograph of Orion "not only superseded all previously existing delineations of that strange object, but virtually prohibited any such being attempted in the future. Changes in its condition, it was made plain, must thenceforth be investigated by a comparison of photographs."[114]

The dream of Charles Piazzi Smyth and his contemporaries, of making drawings accurate enough to be valuable decades or even centuries later, lay in tatters. The idea of using pictures as historical records of the sky was not abandoned: as the description of Common's work makes clear, astronomers still thought of accurate imaging as a way of detecting changes in the appearance of astronomical objects, but now photographs would provide those records. Isaac Roberts thought Edward Holden's moon photographs showed excellent detail, and told him in 1890 that they would "indicate any considerable change upon it."[115]

PHOTOGRAPHY AND FIELDWORK

All these developments affected eclipse fieldwork, and the structure of British expedition planning magnified their impact. Both the Royal Society and RAS created permanent planning groups in the 1880s (which had overlapping memberships), and in 1894 they joined forces to form the Joint Permanent Eclipse Committee (JPEC). The JPEC had a virtual monopoly on government funds and favors, control over party membership and observing agendas, and the power to buy its own instruments.[116] Many of its members had a strong interest in photography and the design and construction of large instruments. Hugh Newall, William Huggins, and Frank McClean were innovators in astrophotography; Herbert Hall Turner (the son of a portrait photographer) participated in the Carte du Ciel project; Andrew Common built large photographic reflectors; William Abney championed the military uses of photography and worked on astrophotographic plate sensitivity; and Edward Knobel was manager of the Ilford photographic works. This was a group who put much faith in the authority of photographic records.

JPEC's big instruments. The JPEC imposed a greater continuity in instruments, observing routines, and observers between one eclipse and the next. The JPEC inherited a sizable inventory of instruments, some of which had been in continuous use since the early 1880s.[117] It also bought or subsidized custom-made spectrographs and coronagraphs, along with instrument mounts, coelostats, and chronometers.[118] Two of its cameras were identical twins, designed to be operated by parties as far apart as possible, to produce standardized records that would capture changes in the appearance of the corona. Each was built around a pair of Dallmeyer doublet lenses of four inches aperture and five feet focal length. A lunar image six-tenths of an inch in diameter was recorded on plates with exposures of one to twenty seconds. The cameras were used nine times between 1882 and 1900.[119] The Thompson photoheliograph had an aperture of nine inches and a focal length of nine feet, and it recorded a lunar image four inches in diameter on 12 x 10–inch plates.[120] The JPEC also recycled instructions from previous expeditions, imposed continuity in mounting and focusing methods and exposure times, and repeatedly nominated the same astronomers to go into the field.

Despite these attempts at standardization of machines and practices, however, the instruments' complexity, their status as customized devices, and the uncertainties of field conditions all kept them from becoming unproblematic, self-registering devices. One astronomer described his instruments as "not complete machines such as are turned out by our best opticians, but rather combinations of laboratory and observatory apparatuses."[121] These instruments were more powerful than any used before. They were also larger, heavier, and more difficult to focus and manipulate. Stable equatorial mounts were too heavy to carry into the field, and the instruments were too bulky to be easily used when placed on mounts, so most were mounted horizontally and the sun's image was reflected off a coelostat.[122]

Big instruments and fieldwork. The JPEC's instruments and the policies behind them dramatically changed eclipse fieldwork. JPEC instructions focused entirely on photographic and spectrographic observation, and drawing was left to sailors, soldiers, and volunteers working without official guidance or sanction. Draftsmen lacked the artistic and scientific credentials of their predecessors, and included no members of the Royal Academy, RAS, or Royal Society. Few tried to draw the corona in detail; instead, they noted the shape and extent of faint streamers and the outer corona, which cameras could not record.[123] Photographs of parties in the 1860s and 1870s showed groups of observers outfitted with small cameras and telescopes, each man making his own observations. Large coronagraphs made small instruments obsolete and centralized labor in the field, turning formerly independent volunteers into mechanical assistants.

Finances demanded that the number of observers sent from London remain rather small, but as instruments grew the demand for mechanically competent volunteers mushroomed. Army and navy personnel became indispensable for these small parties, both before and during totality. In the late 1880s and early 1890s, eclipse planners developed a preference for military and naval assistance. The Royal Society's eclipse committee asked the Admiralty to provide "as large a ship as possible for the service of the Expedition" for the eclipse of 1886. Commercial transportation was readily available, but the ships were needed as sources of "skilled artisan assistance and assistance in the observations, not mere means of conveyance."[124] A decade later Norman Lockyer used the same argument to get his own man-of-war. "So much assistance is required for the proper manipulation of these instruments," he told the JPEC, "that unless a strong staff from a Man of War is available I am afraid that we shall not be able to undertake the work at all."[125] Officers and men detailed to assist eclipse parties helped build observatories and darkrooms, called the time during totality, and served as timekeepers, recorders, and photographic assistants. Lockyer, who was especially aggressive in his pursuit of naval assistance, was certain that ships' crews were improved by the experience. On one ship, he claimed, "there had not been a single punishment for six weeks" while it was detailed to scientific work, clear proof that the crew had greatly enjoyed the change of routine. (And a guarantee of future access to naval resources: Lockyer boasted to Ralph Copeland that "the Admiralty will do anything I want *as it does the ships good*.")[126] Lockyer appears to have enjoyed the experience just as much: after the eclipse of 1898, he declared, "It is you, the officers and men of the H.M.S. *Melpomene*, who have been running this camp, and we three [scientists] have just stood by."[127] Despite Lockyer's thanks, in this regime can be seen the de-skilling of independent military observers, and the harnessing of their labor to the cause of instrumental observation. Twenty years before, volunteer officers would have made their own observations; now they changed plates and kept time.

The experience of observation. The evolution of large instruments and team-based fieldwork changed not only the scale of fieldwork. It seems to have changed the experience of observation as well, transforming it into a kind of mechanical performance, more dependent on discipline and self-denial.

There was always a difference between scientific and casual observation of eclipses. "Marvelous [astronomical] effects cannot of course be dwelt upon and enjoyed by scientific men," amateur J. J. Aubertin wrote in 1870, "for their whole attention must be engrossed by their observations and their instruments."[128] But at the end of the century, it seems that older astronomers concluded that the nature of observation itself had changed.

Traces of the shift can be seen in the ways scientists wrote about their work around the turn of the century, like the letter Oxford astronomer Herbert Hall Turner wrote to Norman Lockyer in 1898. Turner was preparing an address on the recent eclipse at the British Association's annual meeting, and was thinking about discussing the difficulties of using bluejackets and others as assistants in modern eclipse observation. He thought astronomers should adopt a more strictly scientific regime: "[To] work in the way you approve is false economy," he told Lockyer. "I think we want the actual observers (or workers, for the "observation" comes now to very little) to be scientific men in the front rank."[129] Ralph Copeland reported that "as regards *seeing* the eclipse" of 1898, he considered himself fortunate: he was able to look directly at the corona for ten seconds while exposing plates.[130] "Workers" rather than astronomers, "observing" rather than seeing; the wordplay based on the tension between the old and new meanings of these astronomical keywords suggests a profound shift in the character of astrophysical fieldwork and the relationship between the observer and the eclipse, and suggests that astronomers were well aware of the costs of mechanizing observation and scaling up instruments.[131]

However, the words of astronomers also contain a measure of pride in their willing submission to this regime of mechanized self-denial, and contempt for someone who actually looked at an eclipse and drew what he saw. Each man had designed expeditions around large instruments, but they wrote with a bit of nostalgia for the days when observers actually *looked* at the eclipse.[132] Assertions of emotional self-control were another part of the shift. Lurid descriptions of observers' emotions were stock elements of eclipse narratives of the 1860s and 1870s, but they disappeared in the 1880s. Finally, eclipse fieldwork became described as an exercise in self-denial. Astronomers traveled thousands of miles, worked for weeks to build field observatories and set up instruments, then during totality turned their backs on the most spectacular event in the natural world.[133] Little wonder that drawing, which depended on individual ability, judgment, and close involvement with the subject, suffered among JPEC members.[134]

THE BRITISH ASTRONOMICAL ASSOCIATION'S
DRAWING PROGRAM

Others were not so willing to give up on drawing. The British Astronomical Association, which sponsored four expeditions for amateur astronomers, attempted to reform eclipse drawing by standardizing drawing styles, materials, and observing practices. Its work was intended to demonstrate the continued relevance of amateur astronomers, and to establish itself as their arbiter and regulator. What is most interesting is that the program borrowed the methods of composite drawing from the artist's studio and transferred them into the field. BAA observers were organized into groups of four. The use of teams of observers was a well-established part of eclipse routine. By dividing work among several people, it was possible to work more quickly and with less distraction, and it was probably also easier to concentrate if one had only a single task. Spectroscopic observations were often made by groups, with one observer guiding the telescope, a second observing the spectrum of the corona, and a third person recording the observers' comments. Visual and spectroscopic observations could also be performed jointly.[135]

In the BAA system members concentrated on one quadrant of the corona, allowing them to observe more closely and "delineate the form and structure of the corona as they actually saw it" rather than simply "suggest the general effect."[136] Sketches were made with white chalk on black paper, and on a common scale. Before eclipses groups drilled, "observing" a drawing of a corona suspended at a height corresponding to that of the eclipsed sun and uncovered for the duration of totality. "After a few nights of steady practice," BAA member A. Keatley Moore wrote, groups "drew closely together, nervousness disappeared, and the combined results became excellent."[137] Directly after totality, while the experience was still fresh in their minds, the group made a composite picture based on the four sketches and including details they observed but did not have time to draw. Finished sketches had been made in the field before, but the use of group witnessing as a guarantee of their credibility shows that the BAA owed a debt to the methods developed for making composites.[138]

These efforts could happen in part because drawing was pushed to the margins of official expeditions in the 1890s. The BAA's program represented an attempt to carve out a place for amateur astronomers in eclipse observation by developing a response to the constraints of eclipse fieldwork that did not involve large instruments, an exclusive reliance on photography, and a mechanical style of fieldwork. It did not succeed. The association's reports reproduced some interesting drawings, but by 1905 the BAA's parties had devolved from teams who "drew closely together" back into groups of individual observers who made no pretensions of being able to compete with professional astronomers.

Photomechanical Reproduction of Eclipse Photographs

Drawing was driven out of eclipse fieldwork by a combination of pressures: professionalization, changes in instrumentation, and a more basic change in the experience of (and definition of what constituted) scientific observation. These forces could not help but affect the way astronomers made choices about how to reproduce and publish photographs, and in particular, to increase suspicion against composite drawings. Indeed, the period saw a growing distrust of obvious human intervention in the reproduction process.[139] At the same time, new photomechanical reproduction technologies, particularly the relief halftone and photogravure, made it easier to copy photographs directly onto the printed page without the intervention of an engraver. Both held out the promise of permitting direct mechanical reproduction of mechanically produced pictures, of keeping images sealed off from human hands in a closed circuit of chemicals, glass, and steel.

HALFTONE PRINTING

Halftone printing began with the making of a negative of the image to be printed. Negatives were normally made on wet plates, mixed by the photographer according to his own personal (and often secret) formula. Dry plates were faster to use, but were not "as rich in contrast, transparency, and sharpness."[140] The character of negatives was further determined by the halftone screen and diaphragm. The shape and size of the diaphragm determined the negative's sharpness, intensity of contrasts, and depth of shadows.[141] Contrasts were determined by the size of the halftone dots and the degree of their connection; as long as you didn't look too closely, you would not see the dots, but fields of grays or blacks.

Once the negative was developed, it was stripped off the glass plate and put back on reversed. Meanwhile, an enameled copper plate was warmed over a Bunsen burner until it turned brown. The negative and copper plate were then set together in a wooden frame, and set underneath a window or swinging lamp. The areas of the copper plate exposed to the light through the negative would then begin to harden. Because the copper plate was underneath a negative, the dark spaces in the photograph—that is, the halftone dots—were light, and vice versa.

After several minutes the copper plate was removed from the frame and the soft gelatin washed away.[142] The plate was allowed to dry, covered with ink and a dark red vegetable resin called dragon's blood, and then heated. The dragon's blood melted and fused to the hard gelatin, forming an acid-resistant layer that slowed the etching process. The other side was coated with asphalt. The plate was then immersed in a tray of acid. The acid bit first into the most exposed areas of the plate (the lightest parts of the picture), while the areas under the gelatin were protected. As the dragon's blood was eaten away it made the acid red and cloudy. The etching's progress was measured by scratching off a small area along the plate's edge and feeling the depth of the bite with a fingernail.[143]

When the acid had done its work, the plate was removed from the bath, rinsed, and cleaned, and the asphalt taken off the back. The result was referred to by printers as a "flat etching." A proof was drawn and examined closely, and corrections made by hand. After the plate was finished, it was trimmed and mounted on a piece of wood.

PHOTOGRAVURE

The other major reproductive technique used in printing astronomical photographs was the photogravure. Photogravure dominated the market in high-quality art reproductions and was widely regarded as the finest and most demanding printing method ever created, unequaled for "softness, richness of ink impression, lack of visible screen"—a particularly important quality for astronomers, given their desire for images to appear as if they had traveled from eyepiece to printed page—and "enormous capacity . . . for smoothly graduated chiaroscuro."[144]

Photogravures began with a glass positive on "collodion, on a gelatin dry plate, or on carbon . . . according to the subject and the quality of the original negative." A second glass plate, coated with light-sensitive gelatin, was attached to the positive, and an exposure was made. The gelatin turned hard and water-resistant in exact proportion to the intensity of the exposure. It was then peeled off the glass and affixed to a sheet of copper with an aquatint ground and an asphalt back. An intaglio image was made by immersing the plate in a mordant. "The action of the bath commences first . . . [on] the shadows or darks of the picture," photogravure expert Ernest Edwards explained. "The lights are protected from the action of the chloride, the gelatin covering them having been water-proofed by light, whilst the half-tones being partly water-proofed are partly protected." After the acid etching was complete, the plate was dried, the varnish removed from the back, and the plate was steel-faced by "depositing an infinitely thin layer of steel all over the surface of the plate, thick enough to preserve the copper from injury, but thin enough to retain the very finest gradations."[145]

SURVIVAL OF COMPOSITE DRAWINGS

Halftone and photogravure made it easier to mechanically reproduce photographs, but they alone did not make composite drawings, or other forms of reproduction involving artists, obsolete. The RAS *Monthly Notices* and *Memoirs*, and the Royal Society *Philosophical Transactions* and *Proceedings,* published halftone and gravure illustrations from the 1880s, and until 1900 drawings were reproduced mechanically as often as photographs.[146]

The 1889 *Philosophical Transactions*, which published observations of the eclipses of 1883 and 1886, were printed during this transitional phase. Both articles were illustrated with plates drawn by William Wesley. The drawing of the 1886 eclipse was a composite made from a dozen negatives. For the 1883 eclipse, however, Wesley was ordered to make *two* drawings, one of the inner corona "taken from the photographs which had but short exposure," the other "sketched from the photographs to which long exposure had been given" and showing the outer corona. All three drawings were printed as photogravures.[147] These were Wesley's last published eclipse drawings for the Royal Society or RAS. He drew other composites up until 1900, but they were published by the BAA or used in popular lectures.[148] It took several years for the idea that *only* photographs should be reproduced photomechanically to become established.[149] (The question of mechanical reproducibility was, it seems, also affecting drawing. An astronomer in 1892 warned against the use of pencil drawing: "[A] special drawback to its employment [is] its being badly, if at all, suited for [photomechanical] reproduction."$)^{150}$

In the 1890s mechanical reproduction became the rule in JPEC publications, as astronomers tried to complete the circuit connecting coronagraphs with photogravures, and connecting images produced in the field to readers. The photographs that large cameras produced were more detailed and impressive than those produced in the 1870s, and improvements in plate holders and exposure times meant that there were more of them. In their reproduction, intermediaries like engravers were (so it seemed) cut out, and all intervention was avoided. Photographs were not even retouched before printing, but reproduced with all their flaws intact. The plates accompanying eclipse accounts looked impressive at first, but on closer inspection readers could see dark spots from dust, light spots from overexposure, and irregular skies from halation.[151]

The shift away from composites was not motivated by a desire, as George Airy and James Tennant wanted decades earlier, to present readers with the scientific facts of the case, but rather by a distrust of artistic judgment and interpretation. Authors selected the best plates for illustrations, and made no effort to reproduce all of them.[152] Further, astronomers knew that photomechanical processes would not show all the faint and fine details of originals, and that in renouncing intervention they were embracing imperfection. Plates carried captions warning readers of flaws, the absence of faint extensions that could be seen in the negatives, or the failure of the original to record the entire phenomenon.[153] One article even published two different pictures from the same negative, varying the exposure times of each plate to bring out different details visible in the originals.[154] But as unsatisfying as this regime might have been, retouching to adjust light contrasts and bring out those details was now far worse.

COMPLICATIONS OF REALISM

The warnings included in captions and explanations of plates are but one set of indications that the reality of photomechanical reproduction was far more complicated than the idea. It cut engravers out of the copying process, and gave pictures the appearance of being untouched by human hands. But artisanal skill and judgment were required to give plates just the right appearance of nonintervention. Despite the desires of scientists or managers, high-precision automation technology often relies at certain critical points on human skill.[155] Engravers spoke eloquently about the skill and artistic sensibility required in the age of mechanical printing. "A good plate cannot be produced by accident, nor is it a simple mechanical result," one engraver wrote.

There is much skill, taste, knowledge and talent required to make a good negative as there is to any other kind of work. . . . The machines do not work alone and the fine outfits and choice chemicals do not turn out perfect plates. Skill, practice and a thorough knowledge is required to have a first class result; without these all else is useless.[156]

Even though they worked with such hearty materials as copper plates, asphalt, acids, routing tools, and burnishers, engravers needed sharp eyes and steady hands. Halftone screens with more than three hundred lines per inch required special papers, inks, and "exceptional facilities and skill" to use properly; astronomical illustrations used screens as fine as four hundred lines per inch. Dragon's blood was difficult to spread evenly, and the dusted plate had to be warmed slowly for the powder to melt right. In winter, water vapor could condense on halftone screens and glass plates, making etchings foggy and indistinct. Drawing good proofs of flat etchings was so important that certain pressmen specialized in the work. Retouching was important and legitimate enough to attract specialists and have its own tools, workspace, and manuals.[157]

Photogravures likewise required skilled hands and experienced eyes, for the process was delicate and images were highly plastic until printed. Glass positives could be retouched "to any desired extent" before being copied onto metal plates. Experience and a sharp eye were needed to control the etching process, for the acid worked rapidly and mistakes were "very easily and very quickly made." Retouching had to be done by craftsmen who specialized in photogravure, for the skills learned in halftone or line retouching were not transferable. Finally, the steel-facing had to be done by a "skilled printer who is also an artist."[158]

Danger and opportunity coexisted in etching and proofing. At those points the image was unstable, malleable, and threatened to slip out of control. It could dissolve in acid or under the artist's burnisher, or resolve into sharp lines and delicate shades and contrasts. Craft and artistic skill could be used to change the appearance of images, or to make sure that they did *not* change, that the halftone looked as much like the original drawing or photograph as possible. Choices of screen and diaphragm, exposure times, lighting levels, etching depths, and printing materials could all be made with the purpose of getting the negative to produce an image identical to the original—not of *altering* the image but stabilizing it, guiding it unchanged from negative to copper plate to paper.

This illustrates the central problem in astronomical printing, and one of the critical issues and paradoxes in scientific representation: one could hardly afford to alter an engraving for fear of damaging the image, but intervention was required to stabilize the image. Mechanical reproduction could not be effective without human intervention. The irony, as Wilhelm Cronenberg wrote, was that the engraver had to "avoid working like a machine" for photomechanical printing processes to succeed.[159]

The introduction of halftone and photogravure to astronomical journals came just as large eclipse instruments were enjoying unprecedented success in the field. In the eyes of JPEC members, the forces that had made composite drawing necessary, the constraints that had held their predecessors to a mixed and insecure system of representational technologies, were being swept away by new technologies in and out of the field. Large cameras now produced huge, detailed, and expensive pictures of the corona in the field, while the development of halftone and photogravure printing technology made it possible to reproduce them without the aid, or at least the skilled intervention, of artists and printers. Production problems still remained, but debate between artists and astronomers over the proper appearance of the corona was ended when photomechanical copying eliminated the artist. Multiple witnessing, previously so important for establishing the credibility of pictures, was no longer necessary. A single powerful witness could now be trusted to reveal, through a single photograph, astronomical reality in the field and to the masses.

Conclusion

The transit of images from the field to the printed page was long and complicated. The process was full of opportunities for what sociologists of science would call the "social construction" of scientific work, and indeed there is little about the way astronomers chose to observe eclipses that was not subject to negotiation or change. When it developed as a field of inquiry in the early 1860s, astrophysics was dominated by gentleman observers, who worked in the observatory alone or with small groups of collaborators and assistants. They dealt with the various constraints of eclipse fieldwork with strategies that drew on small instruments borrowed from the observatory, whose performance depended mainly on the skill of individual observers. They were influenced by technologies in the laboratory and printer's studio for analyzing and reproducing pictures, and published eclipse pictures were composites of photographs collaborated upon by astronomers and artists.

Forty years later, everything was different. In astronomy as a whole, amateurs were pushed to the margins, displaced by professionals working as staff members in large observatories, using instruments hundreds of times more powerful than their predecessors. Visual observation and drawing were displaced by photography, which while not perfect, was considered infinitely better and more reliable than the eye. Photomechanical reproduction processes consolidated photography's new dominance, by making it possible—in principle, if not always in fact—to copy photographs exactly without the intervention of engravers or artists. Eclipse expeditions reflected these changes. Expeditions in 1900 were directed by professional astronomers, used immense custom-made instruments, and relied for their success on the collective mechanical aptitude of teams of observers, not the sharp eyes and quick hands of individuals. Astronomers depended almost exclusively on photography, and reproduced their photographs using processes designed (imperfectly) to exclude artists. Even the ways astronomers talked about eclipse fieldwork, and the meaning of astronomical keywords such as "seeing" and "observation," changed as instruments grew larger, observing plans more complex, and the number of people required to operate an instrument increased.

In this history one sees many familiar patterns. The professionalization of the scientific community, the increasing trust in photography and other forms of mechanical recording, the scaling-up of scientific instruments, the growing importance of government support for science—all shaped the way astronomers made and reproduced images of solar eclipses. But these factors express themselves in a particular way, in the strategies astronomers developed to deal with the fact that total eclipses are very short, impossible to see anywhere but in restricted locations, and very expensive to observe. Astronomers dealt with the constraints of the eclipse, first in the creation of observing programs and choice of instruments before expeditions, then in the fieldwork itself, and finally in the ways they made sense of visual records generated during totality. The constraints remained the same, but the solutions varied over time, and in those solutions one can see the interplay of social and cultural forces on science.

CHAPTER 5

Astrophysics and Imperialism

Throughout this book, my examination of eclipse expeditions and observations has centered on practices and experiences: what the work of observing eclipses was, and what it was like. The availability of detailed descriptions of travel, camps, fieldwork, instruments, observations, and comparisons of British and colonials' behavior made the choice seem eminently reasonable. This choice is not self-evident, however, nor does it cover everything. But it provided a justification for examining planning, made it necessary to make sense of accounts of native behavior, and provided a reason for writing about what travel and fieldwork felt like to Victorian scientists. It also revealed connections between activities, technologies, and phenomena that might otherwise have gone undetected. The effects of the scaling-up of instruments and the mechanization of observation on the way astronomers felt about eclipse fieldwork would have been difficult to detect if descriptions of rehearsals and the work of observation were read differently. Likewise, the sometimes intimate relationship between the instruments, field recording practices, and copying methods that produced and reproduced pictures of the corona would almost certainly have escaped notice.

This chapter uses the prism of practice to explore the relationship between the material culture of imperialism and the conduct of Victorian scientific fieldwork. My argument is extremely simple: imperialism made eclipse fieldwork possible. This is not to argue a simple technological determinism. It should be self-evident that imperial technologies did not "lead to" or "cause" eclipse expeditions. But expedition planners, colonial volunteers, and astronomers all took for granted that certain technologies, services, skills, labor, and power had to be available in the field for expeditions to succeed. They chose field stations on the basis of the presence of those resources, and those choices had a powerful effect on the culture of expeditions and the nature of observation. This was so because astrophysics, as I argued in Chapters 3 and 4, was a high-tech, high-precision discipline. One essential feature of its character was that spaces—technical and social spaces—were profoundly important to it, and had to be ordered and controlled for it to succeed. Astrophysical observatories were tightly regulated, both socially and physically. An unruly or poorly trained assistant, vibrations transmitted into a telescope's pier, temperature changes expanding or contracting a prism, and shortages of chemicals or other materials—all presented equal challenges to astrophysical order, and all were managed by controlling the spaces around the instrument. If one accepts the idea that technologies and artifacts can be distinguished from one another by their degree of interconnection, and that two intimately connected artifacts blur into each other, then the distinction between the "instrument" and the "observatory" was disappearing in nineteenth-century astrophysics. Different terms were used for the observing devices and the spaces in which they were placed, but one was now so dependent upon the other that the line separating them no longer held any meaning. Astrophysics could not exist outside the space that supported it.

My argument will unfold as follows. I begin with a quick tour of the Victorian observatory. I will not be "opening up the black box" of the observatory so much as describing the design and ecology of a Wardian case, an artificial and self-contained world that permitted a traffic in certain goods and phenomena and prohibited others.[1] I pay particular attention to how the observatory constituted a physical environment and workplace, constructed with the same care and just as necessary to good science as big telescopes and cameras. I contrast the observatory with the field, to show what physical, technical, and human challenges eclipse observers had to overcome in order to make their exceedingly—and as time went on, increasingly—complex observations. Next I analyze the twofold strategy British astronomers used to ensure the stability of their observations. I first examine siting strategies that kept camps within areas brought within the spheres of European (and mainly British) political and technological control. Correspondence showing how field stations were chosen, and maps showing the confluence of railroads, telegraph lines, military forts, and cantonments, weather patterns, and sources of European-trained skilled labor around eclipse camps, will show how British astronomers re-created the observatory as a technical and social space thousands of miles from Greenwich and Cambridge. I then examine the instruments and practices created for and adapted to plein-air solar observation. I show how practices were developed to minimize the effects of nervousness and anxiety brought on by totality, and how the design of specialized instruments reveals an assumption on the part of their makers that they would be used only in European-dominated areas.

Technological sophistication, military power, a culture highly conscious of time and precision—these were all things essential to astrophysical work, and to Victorian astronomers these were things synonymous with Europe and empire.

The Orderly World of the Astrophysical Observatory

The challenges of fieldwork came from a combination of natural conditions and technical constraints. This can be seen by reviewing the process of choosing field sites for expeditions. Even today, geography, time, and weather all have to be factored into account when planning an expedition. Eclipses are visible in a swath of territory a hundred miles wide, but much of a shadow path is cast over ocean, desert, over uninhabited or uninhabitable territory. Few Victorian eclipses were visible in cities, and only one took place over an observatory—the eclipse of 1901, over a small French government facility on Mauritius. Time is an even crueler limiting factor. Eclipses last between two and seven minutes, depending on where you are under the shadow path. Finally, there is the weather. Areas prone to fog, dust, or clouds, or known for regular daily rains, have to be avoided. Out of tens of thousands of square miles, astronomers were lucky to find a few hundred square miles—part of an island here, a dry plain there, some hills farther away—that were close to the center of the shadow path, offered sufficient time for observing the eclipse, and made at least a tentative promise of clear skies.

But the Victorian astronomer's ideal field station offered more than natural advantages: it had to offer an abundance of resources best described as industrial. Astronomers needed clean, potable water; chemicals; secure piers for their instruments; skilled assistants; repair facilities with advanced tools, staffed by skilled machinists; transportation for their fragile, delicate instruments; and a sense of their position in space and time. Astrophysicists needed all these things because of the pedigree of their instruments. Astrophysics and solar physics were born and bred in the observatory, in highly regulated and controlled environments, which both protected sensitive instruments from variations in temperature, humidity, and other natural factors, and connected them to suppliers of chemicals, photographic equipment, and skilled labor. Astrophysics could no more be done outside this supporting environment than it could be done without spectroscopes or cameras. To see why this was so, we need to look at the nineteenth-century laboratory, and in particular at astrophysical observatories.

SPACES OF SCIENCE

The practice of science has long been associated, in one way or another, with specific social or physical spaces. Owen Hannaway, Steve Shapin, and Simon Schaffer have shown that the creation in the early modern period of social spaces for experiment was crucial to the development of natural science.[2] In the nineteenth century, the number and variety of spaces for scientific practice—lecture halls, teaching labs, research labs, and buildings for industrial testing and product development—increased as scientific research and teaching became differentiated and more specialized.[3] Scientists and architects designing laboratories had to deal with constraints imposed by surroundings, the demands of patrons, and the needs of students; but physical considerations dominated plans. More and more, science relied on extremely sensitive instruments that could work only in spaces that were vibration-free, had practically uniform temperature, were shielded from electric and magnetic fields generated by other instruments, or contained pure air and water. Within a building, vibrations from machine shops, heating and cooling equipment, electrical generation, and even student traffic had to be dampened. Outside, streetcar and subway traffic generated additional noise that interfered with precision balances and sensitive instruments. In response, most buildings were designed to be exceptionally stable. Many were designed with steel and iron frames and heavy masonry walls, and all were set on massive concrete foundations. Inside the laboratory, designers made more attempts at insulating instruments from natural and artificially produced noise. Machine shops, generators, and ventilation equipment were often placed on separate foundations or designed to minimize vibration during operation. Special free-standing, isolated piers made of concrete, resting on foundations consisting of combinations of gravel, oil-cloth, and cork, were installed for precision instruments. Optics and spectroscopy laboratories were outfitted with special temperature and light controls.[4] As a result, an architecturally unified exterior could cover a variety of highly differentiated interior spaces. The George Holt Physics Laboratory at University College, Liverpool, designed under the watchful eye of Oliver Lodge, had a steel and concrete frame—incongruously clad in a Gothic red brick and stone exterior—to ensure that it would be "noiseless and steady," and special rooms with temperature and light controls.[5] Likewise, the physics building at Owen College, Manchester, contained two rooms specially designed to house Rowland gratings, a third for temperature-sensitive experiments, and a fourth for electromagnetic research.[6]

Designers also tried to engineer spaces for social purposes. As Charles Baskerville wrote in 1908: "[A] laboratory is a workshop," a phrase that implied for him a need not just to attend to technical issues but also to engage in monitoring and control of workers. In particular, undergraduates had to be kept as far away from research facilities as possible, and their own lecture halls and labs had to be designed so their inhabitants could be easily watched and reprimanded. Doing otherwise, Baskerville warned, invited "the degeneracy of liberty into license."[7] Graduate laboratories, in contrast, were designed with different principles in mind. Arthur Schuster outlined a designer's options: on the one hand, he could "differentiate . . . between rooms and between different classes of students"; on the other, he could design common labs throwing together students of different classes with a variety of interests. German laboratories were designed according to the second principle, and he found they had many advantages. While studying with Helmholtz, he recalled, he shared a laboratory with half a dozen fellow students; as a result, "We became interested in each other's work, and thus increased our experience and gained a much broader view of the range of physics."[8]

THE ASTROPHYSICAL OBSERVATORY

Astrophysical instruments and research had particularly intimate relationships with their observatories.[9] Part of the reason was size. Telescopes were among the largest scientific instruments ever built, and they both required exceptionally large facilities and were exceptionally sensitive to fluctuations in temperature and humidity. Smaller instruments were just as sensitive. Powerful spectroscopes, the workhorses of astrophysics, required exquisite attention and careful use. No two large spectroscopes worked exactly alike, and the focal length was different for different sections of the spectrum—red was longer, blue was shorter—but all spectroscopes were notoriously sensitive to changes in temperature. A few degrees' change in an observing room could affect dispersion and deviation rates enough to require correction, and readings could differ considerably with the seasons. Astrophysicists installing new spectroscopes had to map the focal planes of the colors as temperature varied, and make adjustments to the instrument and their calculations accordingly.[10] Large spectroscopes, like those at Potsdam, Lick, and Yerkes, could even be affected by the heat of an astronomer's hands, and prism boxes had to be insulated in layers of velvet.[11] Spectroscopic observation also took a lot of time. Hours were spent at the telescope focusing and refocusing the spectroscope, bringing one and then another section of the spectrum into view, then reading and checking the positions of lines off a micrometer. The incorporation of photography did not shorten the time involved in making observations, since spectrograms required exposure times of hours or even days, and additional hours were required to measure lines. As spectrographs became more powerful, the number of lines to be measured exploded, creating serious bottlenecks at some institutions. Larger research projects and catalogs of stellar spectra could take years to complete, and many a career was started with work done on plates farmed out from a major research center to faculty at smaller observatories or colleges.

Astrophotography was likewise a time-consuming, demanding affair. Photographs of the moon could be taken quickly, but deep-sky objects could require exposures of several hours, even after the introduction of dry gelatin plates. Consequently, astrophotography required not only good optics and sensitive photographic plates but also an array of tracking, stabilizing, and control technologies. Telescope mounts had to be stable enough to absorb vibrations from sources outside the observatory, from machinery attached to the telescope, and from the movement of the telescope itself. Clock drives, which move the telescope and allow it to stay trained on one part of the sky, were critical elements in the system. The adoption of electrically powered drives made tracking much smoother than previously, and helped make long exposures feasible. Photography changed telescope design, stimulating the development of correcting lenses for refractors, boosting interest in reflectors, and forcing redesign of mountings to compensate for changes in balance created by adding cameras and spectroscopes. Nevertheless, this complex of technologies didn't entirely eliminate humans. Astronomers no longer recorded what they observed through an eyepiece, but they were still needed to make corrections to clock drives.[12] (The irony was that while machines were applied to astronomy to eliminate the personal equation inherent in human observation, astronomers ended up being applied to machines to correct for mechanical errors.)

The sensitivity and precision of astrophysical instruments created a demand for spaces that protected them from temperature variations, damp, vibration, and other distractions. This was not unprecedented in astronomy. Whether devoted to the production of extremely precise positional measurements for use in celestial mechanics, or precise spectroscopic measurements and photographs of deep-sky objects, observatories had long been designed with issues of environmental control and stability in mind. Magnetic instruments, electric clocks, and the metal fittings in transit circles were all sensitive enough to temperature change to require fine climate control. Consequently, heating systems were designed to stabilize ambient temperatures and minimize temperature gradients from one room to another, or between rooms and stairwells or closets. As telescopes became larger, vibrations from street traffic or motors became intolerable. Engineers responded by designing massive, vibration-free piers, often made of granite or stone, resting on foundations independent of the floors and other parts of the observatory.[13]

The astrophysical observatory functioned as a filter, excluding Nature's noise while letting in the signal. It stood at the nexus of a diverse and ever-changing network of instrument-makers, chemical suppliers, journals, scientific societies, and skilled assistants, photographic manufacturers, and electricians, which supplied it with new chemicals, pure metals, fresh batteries, repairs, expert advice, and new tools for prying open the secrets of the heavens. This meant that the astrophysical observatory evolved into a very different kind of space than that of the more traditional astronomical observatory, a workplace devoted to precise measurement of stellar position and motion. William Huggins described the scene at his Tulse Hill observatory in the 1860s when he first became interested in using spectroscopy to study the chemical composition of stars. Not only did his research interests change dramatically, but the observatory itself began to mutate into a deeply industrialized space. As he recalled, Tulse Hill took on

the appearance of a laboratory. Primary batteries, giving forth noxious gases, were arranged outside one of the windows; a large induction coil stood mounted on a stand on wheels so as to follow the positions of the eye end of the telescope, together with a battery of several Leyden jars; shelves with Bunsen burners, vacuum tubes, and bottles of chemicals, especially of specimens of pure metals, lined its walls.[14]

Astrophysics was a hybrid discipline, and astrophysical observatories like Tulse Hill reflected the discipline's multiple origins. Astronomical photography required glass plates, silver compounds, developing baths, and when undertaking especially difficult projects, the expert advice of commercial photographers. While mapping the spectra of stars, astrophysicists borrowed equipment from chemists, opticians, and electrical engineers. As a result, the observatory had to combine elements of the traditional astronomical observatory with the chemical laboratory, physics laboratory, studio, and machine shop, and this heterogeneity required the development of new industrial connections and managerial skills. This raised both theoretical and practical difficulties. As Simon Schaffer has argued, the variety of astrophysics' connections with other disciplines and practices, and the speed with which it could appropriate instruments and techniques from other fields, was both a great strength and a serious weakness. It made the discipline fast-moving, but for figures like the authoritarian George Airy, it also raised questions about the reliability of individual results and its overall legitimacy as a science.[15] Astrophysical instruments' dependence on constant influxes of chemicals and glass plates, protection from environmental distractions, and careful manipulation made their transfer into the field a difficult matter. Even the most disorganized observatory, with pieces of instruments scattered about and Leyden jars lining the walls, was a stronghold of order and stability in contrast to the field.

Imperialism and Technological Systems

It was impossible to do astrophysics without the protective and supportive environment of the observatory. Instruments were too dependent on buildings that shielded them from the vagaries of the environment; research projects relied too heavily on the supplies and skills that could be brought together in the observatory; and the discipline needed stable alignments of technologies and social interests to answer concerns about the reliability of its results. This would seem to make astrophysical fieldwork an impossibility, and almost everywhere outside the observatory and under the path of totality, it was. But it was not completely impossible. In the late nineteenth century, an observatory could be rebuilt in the field, reconstructing both the physical and the social networks that supported spectroscopic and photographic investigations of the sun. Submerged beneath the exotic landscapes on which parties made camp were technological systems that provided transportation, time-keeping, and communications; guaranteed the presence of skilled assistants and repair facilities; ensured access to military and police protection; and promised other services that kept expeditions running and connected to other observing stations and the rest of the world. Just as astrophysics was a high-tech, industrialized science, so too was nineteenth-century imperialism. To fully understand how astrophysical fieldwork was possible, we need to look more closely at the technological side of imperialism, to see how it worked, what impact it had on colonial territories, and how it shaped scientific fieldwork.

Technology has long been recognized as a critical factor in making possible the remarkable expansion of European empires in the nineteenth century. As Daniel Headrick showed in his classic *Tools of empire*, new technologies such as the repeating rifle, gunboat, and quinine prophylaxis made it possible for small numbers of European soldiers to defeat well-organized and trained armies, project naval power into inland Africa and China, and turn malaria from an inevitably fatal disease into a chronic but manageable danger. Likewise, colonial officials saw technological systems like the telegraph, the railroad, plantations, and factories as essential for developing and controlling colonies.[16] The introduction of these technologies and industries had a profound effect on colonial environments. In fact, Alfred Crosby argued, the history of imperialism is the history of ecological change. His *Ecological imperialism* rewrote the history of European settlement of the temperate Americas and Antipodes in terms of the triumph of the European "portmanteau biota" over native ecologies, an interlocking complex of weeds, food crops, animals, and diseases that settled with farmers and herders in the Argentine pampas, New England, New Zealand, and elsewhere.[17] Even if one does not accept Crosby's contention that weeds and animals were more significant actors in the history of colonialism than missionaries, conquistadors, and capitalists, one must recognize that the creation of plantations, transfer of plants from one continent to another, clearing of land for cattle and sheep, opening of timber operations, building of transportation infrastructures and cities, and efforts at game management and conservation—in short, some of the central activities of nineteenth- and twentieth-century colonialism—can be seen as subsets of the history of ecological change, the unconscious and conscious effort to "Europeanize" lands by making them more attractive to European capital, production, and people. They also have to be taken into account when writing the history of colonial science and scientific fieldwork, for they have a decisive impact on what constitutes the field.

RAILROADS, PLANTATIONS AND RANCHES, AND ECOLOGICAL CHANGE

Technological systems played a critical role in the nineteenth-century Europeanization of colonies, and none was more important than the railroad. It was essential to getting valuable but bulky commodities to refineries, mills, and factories, and thence to markets and ports. Prerailroad transportation from plantations and forests in the highlands or interiors to cities and ports was often erratic, slow, and prohibitively expensive. Getting goods to market over narrow dirt roads was a major undertaking, and a large part of a farmer's profits could be eaten up by accidents, spoilage, high transport rates, and delays caused by washed-out roads and bridges. This in turn put limits on the extent of cultivation and herding, keeping them relatively close to cities and markets. Colonial agriculturalists looking for expansion opportunities could be aggressive advocates of railroad and infrastructure building (and low freight rates). Because of their lobbying and investment, rail lines were laid to existing plantations or designed to facilitate farmers' moving into undeveloped hinterlands and hill country. In Ceylon, the coffee planters' Agricultural Association conducted research on cultivation methods and pressured the government for railroads, roads, and ports.[18] Java's railroad system was built to serve foresters, miners, and sugar plantation owners; the last group owned sixty-five hundred kilometers of track by 1930.[19] Regardless of the continent or crop, the basic dynamic between railroads, commercial agriculture, and local ecologies was the same. Railroads brought down the cost of freight transport, made what was an erratic and necessarily seasonal affair routine, and reduced accident losses. With increased transportation capacity, farmers and ranchers were able to expand production, pushing into new areas. This in turn opened formerly untouched fields and forests to developers, who brought them under the plow and hoof, replacing local ecologies with monocultural systems. Sometimes this was done carefully and slowly, but more often a variety of pressures—low economic costs, land tenure laws, local politics, or global commodities prices—conspired to encourage a cycle of rapid clearing, quick planting, and speculation that led to fast exhaustion of newly cleared lands.

This dynamic can be seen more clearly by looking at several nineteenth-century cases. In Argentina, railroad mileage jumped from a few kilometers in 1860 to sixteen thousand kilometers in 1900. This helped turn Buenos Aires into a center of Spanish American culture and finance, but the changes in the hinterlands were even more profound. Frontier expansion had stalled at the Bahia Blanca in the 1850s, and shipment of bulk goods from the interior to the coast was prohibitively expensive and slow. Most agricultural production in the interior was sold in local markets, and ranching was a modest enterprise confined to a region just south of Buenos Aires and oriented mainly toward leather production. With the coming of the railroads, expanded meat-packing facilities, and refrigerator ships, however, Argentine beef could reach European markets, just as demand for meat was on the rise and Continental production was falling. As a result, ranchers pushed their sheep and cattle hundreds of miles south and west into Patagonia, driving onto grasslands that had previously been unaffected by human action. (Linkage of South American cattle to European markets and tastes also encouraged genetic streamlining, as local longhorn breeds were replaced by shorthorns and Herfords imported from England.) The same technological and economic forces remade Canadian ranching at the same time: the construction of railway lines from the prairies to metropolitan centers connected western ranchers to markets on the East Coast and Europe, allowed them to occupy more distant frontier lands that had previously gone undeveloped, and encouraged importation of new genetic stocks. As David Breen observed, the expansion of the ranching frontier owed much to technological developments and was "as much an expansion into the working class areas of London, Manchester, and cities on the American seaboard as into the grasslands of the north-western plains." This was true for Patagonia as well as Manitoba.[20]

The same things happened to crops like wheat and cotton. The Argentine wheat industry was transformed two decades after ranching. Again, vastly reduced shipping costs, increased reliability in transportation, and foreign demand helped turn a domestic crop into an international commodity. As a result, cultivation expanded to 2 million hectares by 1914, the harvests carried first by rail to Buenos Aires and then to Europe.[21] Indian cotton likewise got an important boost from the railroad. Before the railroad reached plantations, cotton had been carried from village to town on unpaved roads, and arrived at market "impure within and dirty without"; after railway lines connected "every cotton field and the sea coast," the commodity was delivered "safely, cheaply, and in an undamaged condition to the port of shipment."[22] The same story played out on the American Great Plains. The railroad made it possible to transport grain by rail from field to city without losing a substantial portion of a shipment to sprouting, mud, and half a dozen varieties of vermin, fungus, and rot.[23]

A few hundred miles north of Argentina, in the Brazilian states of Rio de Janeiro and São Paulo, a similar dynamic between technological systems, native ecologies, and cities and plantations is visible. From midcentury, coffee planters financed railroad projects, ensuring close coordination between rail and plantation expansion. (Earlier some planters had switched from sugar to coffee because of the high losses suffered transporting sugarcane over old roads, while others had chosen tea because it was relatively lightweight and high in value.) In São Paulo, more than seven thousand kilometers of railroad were laid between 1860 and 1930, allowing plantation owners to open up new tracts and expand production: the amount of land brought under cultivation by coffee planters increased *fiftyfold* in the same period. Much of the new farmland had been part of the immense Atlantic Forest, a network of tropical and temperate forests that once stretched thousands of kilometers along the coast of Brazil. To the north in Rio, seventy-two hundred square kilometers of primary forest—18 percent of the state— were cleared for coffee between 1788 and 1888, and another seventy-five hundred for sugarcane. As Warren Dean explains, coffee is especially responsive to careful management and improvements—it can be grown in dense stands, and individual trees can be productive for thirty years—but because of land use practices and rapid changes in commodity prices, coffee in Brazil was grown with far less care than elsewhere in the world. Planters believed that to maximize yields coffee had to be planted on hillsides recently cleared of virgin forest. Hills were cleared by burning to increase soil fertility, and the trees planted in straight lines up the hill, "on the most disastrous master plan imaginable," for laying trees in straight rows maximized runoff and soil erosion. In other areas, coffee was grown in shaded areas and in dense concentrations; in Brazil it was planted sparsely and in the open, reducing yields per tree and per hectare, forcing planters to clear yet more forest to maintain profits. Even though the coffee tree produces for thirty years, groves were abandoned after four, "and new swaths of primary forest were then cleared to maintain production. Thus coffee marched across the highlands, generation by generation, leaving nothing in its wake but denuded hills."[24]

CONCEPTS OF NATURE

International commodity markets, transferred technologies, legal policies, and agricultural practices all had their effect on colonial ecologies. But the history of ecological imperialism involves more than mechanical and large-scale forces: it is also a history of conflicting ideas about nature, animals, and the structure of colonial social relations. Consider hunting in Africa and India. On the European frontier in Africa, hunting was often a major source of food, and a good source of hard cash: mixed with a little farming, a little mining, and trade with Africans, it made the difference to many a settler living hard on the margins. With further settlement and economic development, however, hunting changed. Farmers and herders went to war against native animals (especially predators), clearing land for expensive, imported herds whose products would eventually be sold in commodities markets in Europe. As big game became scarcer and hunting receded in importance as a source of food and cash, it became more of a ritual and tourist activity. The structure of safaris, with their European leaders, African trackers, and large party of bearers and servants, neatly replicated patterns of social and labor relations in the garrison and plantation. As a result, William Storey argues, "the basic underlying structures of the hunt symbolized the triumph of culture over nature and of the colonist over the colonized." Regulation of game by wardens, the creation of rules prescribing the ways game could be killed, and the development of a hunter's ethic that stressed quick kills all worked against native African practices. Game policies often favored tourist hunting at the expense of native hunting, while traditional methods of capturing game were outlawed on the grounds that they were inhumane or unsporting. In some areas, the creation of game reserves threatened to deprive some indigenes of traditional sources of food and livelihood. In others, British hunters saw it as an updating of indigenous royal hunting and game management practices that gave them an air of locally recognized authority, akin to efforts to legitimate British rule in India by taking on the trappings of Mughal rule.[25]

The history of colonial forestry likewise shows how different ideas about the meaning and proper use of nature, the value of plants and animals, and the drawing of boundaries between proper or profitable and improper cultivation shaped colonial policies and affected native societies. In India, British foresters clashed with local peoples from Kashmir to Madras over the management and use of the subcontinent's vast forests. Foresters saw timber as the forests' principal commodity, and sought to maximize sustainable yields, exclude others from using forests, and identify and inventory useful crops. The results could be dramatic for those who had lived in and off the forests. Hunter-gatherers were kept off lands on which they had lived, and their access to game animals declined under the pressures of the shikari and policies favoring recreational hunting over hunting for food. Likewise, slash-and-burn and extensive low-tech methods of farming ("jhum" cultivation) in forest and hill country were discouraged by colonial officials. Jhum farming, they argued, was uneconomical compared with plow and plantation agriculture, and it threw agrarian tribal groups into competition with market-oriented timber contractors. After decades

caught between legal constraints on their livelihood and fights with
timbermen, jhum cultivators abandoned independent farming for share-
cropping and wage labor in timber camps. Farmers and herders on the
plains were also affected, though indirectly. Villages on the edge of forests lost customary rights to collect deadwood for fuel, grass and leaves
for feed, and timber for housing. Further afield, the destruction of forests affected rainfall on the open plains, and climate and river patterns.
Finally, artisanal industries such as bamboo weaving and iron smelting
(which depended on charcoal from forests) declined as they were cut off
from the basic raw materials necessary to their trades. In all these cases,
British definitions of what constituted economically valuable forest
products guided policies about what would be extracted from forests,
who would have access to lands, and who would not. Broadly speaking,
Ramachandra Guha and Madhav Gadgil argue, these policies "encouraged industrial exploitation while maintaining restrictions on village
use," toppling common-law practices and customs, and putting intense
pressure on indigenous economies and societies.[26]

RAILROADS, ECONOMIC DEVELOPMENT,

AND MILITARY POWER

All over the world, by lowering transportation costs and losses, plantation owners, farmers, and ranchers were given an incentive to expand
cultivation, pushing back the forest or jungle, driving game and predators out of the area, or throwing native farmers and squatters off the
land. But railroads were more than indirect agents in the ecological
transformation of colonial territories: as major consumers of iron, steel,
and especially timber, they were themselves important ecological actors. In São Paulo, railroad construction and plantation expansion went
together neatly: wood for fuel, railroad cars, and ties was provided by
fazendieros "ready to deforest their own lands for coffee planting."[27]
Late-nineteenth-century estimates for Indian projects give a sense of
how much pressure construction and maintenance of a railroad line
could exert on a colony's forest resources. Even Indian forests, which
covered a quarter of the subcontinent and "abounded in . . . good timber trees," were threatened by railroad building. Engineers preferred to
use sleepers (the wooden ties under the tracks) made of teak, then as
now one of the most valuable and expensive tropical hardwoods. A single mature tree (about thirty years old) could yield three to five ties; a
mile of track used 1,750 ties, or between 350 and 540 trees. By 1868,

5,600 miles of track had been approved for construction, creating a demand for nearly 9.8 million sleepers. Once the system was running, Edward Davidson estimated, another 1 to 1.5 million ties would be needed every year for repairs and maintenance. Engineers in the 1860s thus had to find more than 2 million trees with which to build the railroads, and another 200,000 a year for repairs. This didn't figure in construction of additional lines, which everyone expected would be built. As a result, in a continent full of forests, some engineers expected that they would have to go to the London market for Canadian and American timber.[28]

Railways were also important as strategic resources. The rhetoric of "annihilating distance" had one meaning for a plantation owner speeding tea to market; it meant something else for a regimental commander moving troops from cantonment to frontier. Railroad construction jumped after the Sepoy Rebellion of 1857, in part because the government decided that the railroad was necessary to maintain Britain's hold over the subcontinent. In Sumatra, the first railroad was built to transport soldiers and supplies to the northern province of Achin during the pacification campaign of 1874.[29] But contemporary boosters were most enthusiastic about the railroad's potential for peaceful conquest and development of colonial territories. The *Westminster Review* crowed: "Besides facilitating and cheapening the conveyance of person and goods in all directions, they also disseminate intelligence, knowledge, ideas of all sorts. . . . The effect which a railway may have in contributing to industrial progress . . . is as liable to be over- as underrated."[30] Few writers were so reserved as to underrate the railroad's impact on India, however. There more than anywhere else, it was expected to do great things, to transform a continental economy, to reawaken races caught in Oriental slumber, to break down divisions of caste and class, and to stand as permanent monuments—greater perhaps than laws, schools, courts, or church—to British rule. It was "more powerful" than "laws, roads, bridges, canals, or even education," Captain Edward Davidson wrote in 1868, capable of "overturning prejudices, uprooting habits, and changing customs as tenaciously held and dearly loved almost as life itself."[31] Enthusiast and engineer Rowland Stephenson wrote: "Railways may, with equal force and propriety, be defined as the great modern instrument of civilization—the detached links of that extended chain which, at no distant date, shall connect the remotest ends of the world, and exercise upon the whole family of man a humanizing and irresistible influence." Finally, rail repair facilities were

among the most advanced in a colony, and could serve as a site for dis-
seminating mechanical expertise to native employees. A well-equipped,
two-story shop in a modest junction could be expected to have a dozen
assorted lathes, half a dozen drills, planing and slotting machines, steam
hammers, furnaces and smith's forge, and machinery for precision
work.[32]

Colonialism and Fieldwork

All this explains why eclipse planners consistently preferred areas
served by rail. It was the fastest, most reliable, and most secure form of
transportation available to them, and its presence in a region was a
guarantee that other essential services and technologies would be available even in the field: machine tools in repair shops, skilled assistants,
engineer-informants, telegraph lines, and some of the comforts of home.

RAILROADS AND THE GEOGRAPHY OF
ECLIPSE FIELDWORK

Members of the 1860 eclipse expedition to Spain fanned out along
the Bilbao and Tudela Railway, since by staying close to the line they
could be assured that "assistance, if needed, was at hand." In fact, this
was an area just being brought under the civilizing influence of the railroad. As the Rev. Charles Pritchard recalled, "A few months before our
arrival, but few Englishmen had ever traversed that wood, or perhaps
could have traversed it alone and in safety; but now we wondered at,
and we welcomed as the precursors of civilization, the scream of the
steam-whistle and the screech of the saw-mill."[33]
Because of the close link between railroads, resources, and safety,
planners often limited their search for field sites to railroad stations and
small towns that fell under the path of totality. After taping together his
maps of India and drawing in the path of the 1868 eclipse, John Tennant
circled the towns close to the railroad, and concentrated all further research on them.[34] Thirty years later, the JPEC chose sites for the eclipse of
1898 by the same method. The eclipse was favorably described by Scotland's Astronomer Royal as being "intersected no less than eight times by
one or another of the railways," neatly demonstrating the importance of
railroads to eclipse fieldwork.[35] Astronomers in London and their agents
in the field agreed on the importance of the railroad to the success of an
expedition. Major S. G. Burrard, an Indian Army officer who assisted
William Christie in planning his 1898 expedition, wrote that "in the

Bombay Presidency, there are practically only two possible sites [to make camp], viz. the points where the central line of totality intersects the two railways."³⁶ Thanks to advice like this, three of the four JPEC-sponsored expeditions made camp within a mile of railroad stations, and the fourth made camp at an abandoned coastal fort. Sites accessible by rail were always astronomers' first choice; sites accessible by water were their second. James Tennant, who had carefully researched railroad-supplied towns in India, ended up making camp in Guntoor, a small town near the eastern coast accessible by steamer.³⁷ The presence of railroads in a province could even make the difference between official indifference and enthusiasm. When trying to get money from the India Office to observe the eclipse of 1871, George Airy pointed out that "the whole district is easy of access by railway from Beypoor and Madras." The office yielded, partially subsidizing an expedition.³⁸

What is interesting about all these cases is how prominent a role railroad access played in the choice of field stations. It was not just a great convenience to have a train station close by. Except when observing on the coast, no one thought to make camp anywhere else. By all appearances astronomers took G. O. Trevelyan's wag that civilization in colonized areas extended "one hundred yards on either side of the railway track" seriously.³⁹

TAPPING LOCAL EXPERTISE

Astronomers defined sites that were accessible by rail or steam as the only potentially viable locations for observing eclipses; to make better-informed choices, they had to rely on informants who could provide advice about weather at specific sites, transportation, road quality, hotels, and campsites. As we saw in Chapters 2 and 3, those informants most often were army officers, administrators, and engineers who had their own field experience, were familiar with the quirks of local climate, knew local customs, the going rates for supplies and laborers, and whom to talk to in order to get special treatment in customs or special rates for cargo. As one army officer told Astronomer Royal Christie, he knew a road was well paved because "I have marched it."⁴⁰ Correspondence between local informants suggests that they were perfectly aware of their importance, and felt responsible for the health and safety of their astronomer-charges: a colonel in the Indian Army advising Scottish Astronomer Royal Ralph Copeland was warned by his Edinburgh-educated son that Copeland "is not young and will require to be taken care of!"⁴¹ This kind of assistance was invaluable to astronomers already busy with instruments and observing programs, and would not have been available had parties tried to locate too far from areas known to expatriates.

Local informants could also provide access to local sources of skilled and unskilled labor. It was absolutely essential for the success of expeditions that astronomers be able to recruit labor in the field: after 1870, only astronomers were subsidized by the government, and no party could afford to take along the assistants, instrument-makers, or other "invisible technicians" who kept observatories running and would be needed to keep field camps going. The most dramatic example of railroad staff helping parties came during the eclipse of 1860, when Charles Vignoles, the English-born engineer-in-chief of the Bilbao and Tudela Railway, "converted . . . the whole of the available strength of the establishment, from the managing director and the chief engineer through all the various grades down to the laborers . . . into either astronomers or astronomers' assistants."[42]

This was an extreme example of railroad labor being put to the service of astronomy, but many other expeditions over the next fifty years were joined by one or two station managers, engineers, or administrators. Other parties could not have functioned without European-trained colonial engineers, railroad workers, military officers, and soldiers and sailors. As we saw in Chapter 2, it was routine for local European administrators, doctors, engineers, teachers, and other expatriates to assist with instruments (causing some consternation among nonwhite observers). Military officers and surveyors were especially valuable, since as a Trigonometric Survey general put it, they represented "skilled labor in the shape of observers accustomed to taking accurate observations and to manipulating delicate astronomical instruments."[43] Officers were also valuable because, in addition to possessing the habits of command and exactitude, they could bring subordinates with them: one officer put in charge of preparing a site for the Astronomer Royal commandeered two hundred men to clear the grounds and cut a road to the nearest town.[44] Norman Lockyer's expeditions in 1896, 1898, and 1900 each absorbed the labor of an entire man-of-war; in fact, he broke with the habit of stationing his party along a railroad only because the navy provided him with both transportation and all the labor he needed. This dependence on skilled field labor tightened the connections between expeditions and empire, for those men were most likely to be found leading troops in garrisons, servicing railroads, surveying state borders, and in the government offices of colonial cities.

THE INFLUENCE OF LOCAL CULTURE

As Chapter 3 showed, this dependence also meant that eclipse expeditions were heavily influenced by colonial culture, especially the cultures of tourism and official fieldwork. Most of an expedition was actually spent not in the field, racing against the clock to get instruments up and running and to train assistants, but traveling from Britain to the field and back, and the same technologies and institutions that ferried tourists around the Mediterranean and around the Cape carried astronomers. Before setting out on the high seas and during their weeks in transit, eclipse observers behaved quite a bit like tourists, carefully planning their itineraries, researching destinations, and reading up on local life and customs. While traveling, they rode the same steamers, stayed at the same hotels, and saw many of the same sights; they were separated from their fellow travelers only in being more privileged, moving in more powerful official circles, and being the objects of local elites' curiosity and honors.

Things became more serious when parties reached their destinations. It normally took a few days to get to their camp sites and unpack everything, but once expeditions were settled in, there was an immense amount of work to be done: instruments had to be calibrated or repaired, assistants prepared, and the camp cordoned off from visitors. These camps often borrowed heavily from the cultures of plantation owners, colonial administrators, and military officers. A number of parties simply moved into the guest houses of plantations, government bungalows, or the spare rooms of tax collectors and engineers, while others were outfitted with tents from civil service or army regiments. This gave expeditions an official imprimatur recognizable to both European and native visitors, for the mobile accommodations of officials who spent part of each year "in camp" were just as much emblems of state power as buildings in the capital. The effect was rounded out by the presence of native servants, European volunteers, and the importation of local rituals such as morning hunts. Like the camps of civil servants, these were also workplaces—increasingly busy workplaces. For astronomers, the clock was running down, and there were only precious days or weeks to get everything ready. The last couple of days before an eclipse were a blur of round-the-clock work, especially for astronomers: assistants could be trained to do some things, but all too often custom-made instruments could be repaired or coaxed into performing only by those who had designed and worked intimately with them. As a result, by the day of totality, nerves were often stretched near the breaking point.

INSTRUMENTS, OBSERVING PRACTICES, AND
COLONIAL INFRASTRUCTURES

The preference for sites close to railroads, dependence on local informants, and reliance on skilled labor were as common for expeditions in the 1900s as in the 1860s; but many other things changed dramatically. As I discussed in Chapter 4, two technical developments powerfully affected eclipse fieldwork: instruments grew dramatically larger between 1860 and 1900, and photography displaced drawing as the preferred means of recording the appearance of the solar corona. These changes in turn affected the ways that labor was divided in eclipse parties, and even changed what it meant to make an "observation" of the eclipse. Observers in the 1860s and 1870s worked exclusively with small instruments borrowed from observatories. Useful spectroscopic observations could even be made with hand-held spectroscopes that showed only four or five lines. Drawing was as popular as photography before the 1880s and its records judged as reliable, so long as they were made by engineers, artists, or astronomers who had experience in astronomical or field drawing and were guided by RAS instructions. During totality they would alternate a few seconds' observing with quick sketching, drawing the broad outlines of the corona and some of its more interesting details, and afterward make a second, more detailed drawing that included features they observed but did not have time to record. Photography was considerably more difficult a means of recording the corona's appearance. Photographic plates were slow, and until the 1880s photographers used collodion wet plates, which yielded excellent detail but had to be prepared right before the eclipse and developed immediately after exposure. Officially sanctioned astronomers were in charge of these parties, but they did not control the work of individual observers too closely, and there was little need: this was small-scale science, and talented amateurs could, after a few days' training, work on their own and make observations as good as a professional's.

This system of small instruments, direct observation, and drawing unraveled in the 1880s. Several factors worked to undermine it. First, the JPEC invested heavily in large cameras and spectroscopes designed specifically for eclipse use. These custom-made instruments were hundreds of times more powerful than their predecessors, producing spectrograms with hundreds of lines rather than a handful and coronal photographs a foot rather than an inch square. Second, the division of the astronomical community into professionals and amateurs was replicated in the field. Few amateurs could afford to build instruments that would be used for only a few minutes every two or three years, and the BAA could not match the JPEC's access to government funds and favors. Third, the invention of dry-plate photography, the growth in the early 1880s of a sense that drawing was irredeemably flawed by artistic subjectivity and could never be as accurate or detailed as photography, and the development of new methods for mechanically reproducing photographs pushed drawing to the margins of government-sponsored expeditions. This new system of large instruments profoundly altered the character of eclipse observation in the 1880s and 1890s. Drawing was still used by BAA-organized amateur parties, but on JPEC-sponsored expeditions artists worked without official instructions, and they boasted fewer artistic and scientific achievements than their predecessors. The introduction of faster isochromatic plates in the 1890s, which shortened exposure times for photographs and spectrographs and increased the number of plates that could be taken during totality, and new sliding systems that reduced the vibrations caused by changing plates, increased both the amount of labor required to operate instruments and the speed at which parties had to work. Teams doubled or tripled in size and their labor was divided more finely. As a result, volunteers and artists who had worked independently as professional astronomers' equals in the 1860s were reduced to assistants making adjustments to a knob. Even for professionals, mechanical operation replaced skilled independent work, drill replaced experience, and management of instruments replaced direct observation of the sun. Together, these changes amounted to a revolution in the relationship instruments mediated between observers and the eclipse, documented in the ways that astronomers described their work in the 1890s, especially the shifting meaning of words such as "observation" and "seeing," words central to the astronomer's craft.

While instruments grew larger, and astronomers came to rely more on photography than drawing, they did not grow less dependent on the technological systems of Victorian imperialism. In fact, the scaling-up of instruments and the mechanization of imaging tied expeditions even more closely to railroads, expatriate informants, and local labor. For one thing, the costs of failure increased as the century wound down: no one wanted to risk breaking expensive, one-of-a-kind instruments, and since such devices were larger and more complicated, they were also easier to damage and harder to repair in the field. Eclipse fieldworkers had never been especially adventurous when choosing their stations when they used cheaper instruments; now they had reason to be even more conservative. For another, these instruments required teams of six or eight operators; a well-equipped party in the 1890s could require thirty skilled assistants, and half as many servants and police to keep time, carry plates to the darkroom, and keep the curious (or frightened) at a safe distance. It is significant that even though many of its members were eclipse veterans and knew the challenges of using big instruments in the field, the JPEC made no effort to design instruments that could survive harsh treatment, be operated by a single astronomer, or were "hardened" to survive carriage by pack animals or careless porters. Instruments grew more powerful as scientific tools, but more fragile and labor-intensive, and even more temperamental. They were designed under the assumption that they would be used only in areas in which they could be transported gently, repaired, and operated by teams of assistants: a need for imperial technologies and culture was hard-wired into them.

However, while eclipse expeditions required less time, more equipment, and more local support, they may have differed from other kinds of scientific fieldwork in degree rather than in kind. Even anthropology, a relatively low-tech discipline and the quintessential field science, could profit from the same transformations and technologies that made astrophysical fieldwork possible. This is clear from a 1913 W. H. R. Rivers essay, in which the anthropologist recommended *against* trying to study "people who are as yet wholly untouched by western influence." Natives accustomed to dealing with whites, he explained, were less likely to "fear, or be offended by, inquiry into their customs. . . . Probably the most favorable moment for ethnographical work," he advised, "is from ten to thirty years after a people has been brought under the influence of official and missionary."[45]

Nineteenth-century British astronomers couldn't have escaped the world of Victorian imperialism if they had wanted to: too many choices, from the design of instruments to the selection of camp sites, kept them in the world of railroads, telegraphs, plantations, government bungalows, and regimental headquarters. But no Victorian astronomer had any interest in escaping, or incentive to do so. Even when Norman Lockyer was suffering through the jungles of India in 1871, nearly delirious from opium and heat, he was never so delirious as to try to escape the transportation and communication networks, government offices, services, and protection afforded by the local government: he and every other astronomer relied too heavily on those services to keep them in touch with each other, to complete their own observations, and to ensure that those observations would be reliable. To move outside their realm would have meant jeopardizing the expedition's observations. Too much could go wrong: chronometers might go awry, delicate instruments might be broken, observers would be placed in danger.[46] This territorial conservatism allowed expeditions to be shaped by customs of travel, recreation, and work, customs that could only exist within a complex and specific framework of institutions and technologies. All of this meant that eclipse expeditions, unlike the expeditions of geological surveyors or Antarctic explorers, had nothing to gain and everything to lose by moving into completely unknown territory. Expeditions that sought knowledge about the terrestrial world could succeed only if they moved into new lands; eclipse expeditions, whose "field" was the space in which their work was conducted but whose subject of interest was the sky, could succeed only if they stayed in known lands. Reliable observations of eclipses could not have been made outside the spheres of European civilization and technology.

Victorian astronomers knew this. One of them articulated the unity of science and imperialism that he and his colleagues had forged in an 1871 article published in the popular magazine *Good Words*. Without astronomy's aid, he declared, "the mighty empires now consolidating in the far west and south would not exist . . . and the hope of ultimately linking mankind into one brotherhood of God's children would be abortive."[47] Western and imperial technology may not have linked "mankind into one brotherhood of God's children," but it did link astrophysical workers together, and turn previously inaccessible areas of the world into sites capable of supporting astrophysical research.

Abbreviations

The following abbreviations are used for archival collections:

CAY	Charles Augustus Young Papers, Dartmouth College
DPT	David Todd Papers, Yale University
JNL	Joseph Norman Lockyer Papers, University of Exeter
MLS	Mary Lea Shane Archives of the Lick Observatory, University of California-Santa Cruz
MLT	Mabel Loomis Todd Papers, Yale University
RAS	Royal Astronomical Society Archives, Burlington House, London
RGO	Royal Greenwich Observatory Archives, Cambridge University
ROE	Royal Observatory Edinburgh Archives
RSL	Royal Society of London Archives
SNP	Simon Newcomb Papers, Library of Congress

The following abbreviations are used for journals:

AHR	American Historical Review
BAA	Brtish Astronomical Association
BJHS	British Journal for the History of Science
CSSH	Comparative Studies in Society and History
HSPS	Historical Studies in the Physical and Biological Sciences
JHA	Journal of the History of Astronomy
RASMN	Monthly Notices of the Royal Astronomical Society
SSS	Social Studies of Science

Notes

Chapter 1: Introduction

1. William T. Lynn, *Remarkable eclipses* (London: William Wesley, 1894), 3.

2. I have had greater luck with American sources: see Alex Soojung-Kim Pang, "Gender, culture, and astrophysical fieldwork: Elizabeth Campbell and the Lick Observatory–Crocker eclipse expeditions," *Osiris* 11 (1996), 15–43.

3. Solar physics is the subject of Karl Hufbauer, *Exploring the sun: Solar science since Galileo* (Baltimore: Johns Hopkins University Press, 1991); spectroscopic observation has been treated in William McGucken, *Nineteenth-century spectroscopy: Development of the understanding of spectra 1802–1897* (Baltimore: Johns Hopkins University Press, 1969), and in a very different vein, Simon Schaffer, "Where experiments end: Tabletop trials in Victorian astronomy," in Jed Buchwald, ed., *Scientific practice: Theories and stories of doing physics* (Chicago: University of Chicago Press, 1995), 257–99.

4. There are numerous reviews of this literature; useful recent ones include Jan Golinski, "The theory of practice and the practice of theory: Sociological approaches in the history of science," *Isis* 81 (1990), 492–505; Andrew Pickering, "From science as knowledge to science as practice," in Pickering, ed., *Science as practice and culture* (Chicago: University of Chicago Press, 1992), 1–26.

5. Among the most famous are Bruno Latour and Steve Woolgar, *Laboratory life: The construction of scientific facts* (1979, repr. Princeton: Princeton University Press, 1986); Harry Collins, *Changing order: Replication and induction in scientific practice* (Beverly Hills: Sage, 1985).

6. Martin J. S. Rudwick, *The great Devonian controversy: The shaping of scientific knowledge among gentlemanly specialists* (Chicago: University of Chicago Press, 1985); Steven Shapin and Simon Schaffer. *Leviathan and the air-pump: Hobbes, Boyle, and the experimental life* (Princeton: Princeton University Press, 1985); Timothy Lenoir, "Practice, reason, context: The dialogue between theory and experiment," *Science in context* 2 (1988), 3–22.

7. Steve Shapin, "The Invisible Technician," *American scientist* 77 (1989), 555–63; Pang, "Gender, culture, and astrophysical fieldwork"; Jane Camerini, "Wallace in the field," *Osiris* 11 (1996), 44–65; Stephen Barley and Beth Bechley, "In the backrooms of science: Notes on the work of science technicians," *Work and occupations* 21 (1994), 85–126.

8. The most comprehensive overview is Mario Biagioli, ed., *The science studies reader* (London: Routledge, 1999). As Nicholas Rasmussen put it, re-

cent scholarship shows greater appreciation for the "craft analogy for science implied by the phrase 'social construction'": "Facts, artifacts, and mesosomes: Practicing epistemology with the electron microscope," *Studies in the history and philosophy of science* 24 (1993), 227–65, on 228.

9. See, for example, Andrew Pickering, "Beyond constraint: The temporality of practice and the historicity of knowledge," in Jed Buchwald, ed., *Scientific practice: Theories and stories of doing physics* (Chicago: University of Chicago Press, 1995), 42–55.

10. See Pang, "Visual representation and post-constructivist history of science," *HSPS* 27 (1997), 139–71. Jan Golinski, in his *Making natural knowledge: Constructivism and the history of science* (Cambridge: Cambridge University Press, 1998), uses the term "constructivist" to describe many of the same things I call postconstructivist.

11. There is a large and contentious literature on this issue; two examples (the first useful, the second influential but intellectually useless) are Yves Gingras and Samuel Schweber, "Constraints on construction," *SSS* (1986), 372–86; and Paul Gross and Norman Levitt, *Higher superstition: The academic left and its quarrels with science* (Baltimore: Johns Hopkins University Press, 1994). An interesting response to critics of the Edinburgh School is Barry Barnes, "How not to do the sociology of knowledge," *Annals of scholarship* 8 (1991), 321–35.

12. Carlo Ginzburg, *The cheese and the worms: The cosmos of a sixteenth-century miller*, trans. John and Anne Tedeschi (New York: Penguin 1982); Joan Scott, *The glassworkers of Carmaux: French craftsmen and political action in a nineteenth-century city* (Cambridge: Harvard University Press, 1974). My thinking on this issue has been influenced by Thomas Söderqvist, "Existential projects and existential choice in science: Science biography as an edifying genre," in Richard Yeo and Michael Shortland, eds., *Telling lives in science: Essays on scientific biography* (Cambridge: Cambridge University Press, 1996, 45–84).

13. For another perspective on the sources of popularity of representations, see Michael Lynch and Steve Woolgar, "Introduction: Sociological orientations to representational practice in science," in Lynch and Woolgar, eds., *Representation in scientific practice* (Cambridge: MIT Press, 1990), 1–18.

14. Ibid., quotation on 5; Michael Lynch, "Discipline and the material form of images: An analysis of scientific visibility," *SSS* 15 (1985), 33–67, quotation on 59.

15. Daniel Headrick, *The tools of empire: Technology and European imperialism in the nineteenth century* (Oxford: Oxford University Press, 1981); this was followed by Headrick, *The tentacles of progress: Technology transfer in the age of imperialism, 1850–1940* (Oxford: Oxford University Press, 1988).

16. Lucile H. Brockway, *Science and colonial expansion: The role of the British Royal Botanic Gardens* (New York: Academic Press, 1979); David Mackay, *In the wake of Cook: Exploration, science and empire, 1780–1801* (London: Croon Helm, 1985); Roy Macleod and Philip Rehbock, eds., *Nature*

in its greatest extent: Western science in the Pacific (Honolulu: University of Hawaii, 1988); Anyda Marchant, "Dom Joao's botanical garden," *Hispanic American historical review* 41 (1961), 259–74; Mary Louise Pratt, *Imperial eyes: Travel writing and transculturation* (London: Routledge, 1992); Lewis Pyenson, *Cultural imperialism and exact science: German expansion overseas, 1900–1930* (New York: Peter Lang, 1983); Robert Stafford, *A scientist of empire* (Cambridge: Cambridge University Press, 1989); Susan Sheets-Pyenson, "Cathedrals of science: The development of colonial natural history museums during the late nineteenth century," *History of science* 25 (1986), 279–300.

17. John Cell, "Anglo-Indian medical theory and the origins of segregation in West Africa," *AHR* 90 (1985), 307–35; Philip Curtin, "Medical knowledge and urban planning in tropical Africa," *AHR* 90 (1985), 594–613; Mark Harrison, *Public health in British India: Anglo-Indian preventive medicine, 1859–1914* (Cambridge: Cambridge University Press, 1994); Michael Worboys, "The emergence of tropical medicine: A study in the establishment of a scientific specialty," in G. Lemaine, ed., *Perspectives on the emergence of scientific disciplines* (The Hague, 1976), 76–98. Colonial ecology is the subject of Richard Grove, "Conserving Eden: The (European) East India Companies and their environmental policies on St. Helena, Mauritius and in western India, 1660 to 1854," *CSSH* 35 (1993), 318–51.

18. Edward Said, *Orientalism* (New York: Vintage, 1979); Michael Adas, *Machines as the measure of men: Science, technology, and ideologies of Western dominance* (Ithaca: Cornell University Press, 1989); Bernard S. Cohn, *Colonialism and its form of knowledge: The British in India* (Princeton: Princeton University Press, 1996); Henrika Kuklick, "Contested monuments: The politics of archaeology in Southern Africa," in George Stocking, ed., *Colonial situations* (Madison: University of Wisconsin Press, 1992), 135–69; Gwyn Prins, "But what was the disease? The present state of health and healing in African studies," *Past and present* 124 (1989), 159–79.

19. Deepak Kumar, "The culture of science and colonial culture, India 1820–1920," *BJHS* 29 (1996), 195–209; Satpal Sangwan, "Indian responses to European science and technology, 1757–1857," *BJHS* 21 (1988), 211–32.

Chapter 2: Planning Eclipse Expeditions

1. There are no detailed accounts of the dinner talk at the Freemason's Tavern, or the monthly meeting afterward; readers should take these as logical, but imaginative, reconstructions. On preliminary planning, see Solar Eclipse Committee minutes (RS CMB.2, "Miscellaneous Committees 1869–1884"); Airy to Charles Young, 5 Mar. 1870 (CAY, Box 6, File 6); Royal Astronomical Society circular, 9 Apr. 1870 (RGO 6/131); RAS to E. J. Stone, 19 May 1870 (RGO 6/131). On joint RAS–Royal Society planning, see J. L. E. Dreyer and Herbert H. Turner, eds., *History of the Royal Astronomical Society, 1820–1920* (London: Wheldon and Wesley, 1923, repr. Blackwell Scientific, 1987), 169. Spring and summer proposal-writing is described in Airy to G. H. Richards, 29 Mar. 1870; Richards to Airy, 31 Mar. 1870; Airy to Edward Sabine, 21 June 1870;

Charles Pritchard to Airy, 23 June 1870; Airy to Huggins, 25 June 1870; Airy to Sea Lord, 25 June 1870 (RGO 6/131); the rejection is in Richards to Airy, 4 Aug. 1870; Airy to Richards, 5 Aug. 1870; Airy to De la Rue, 5 Aug. 1870 (RGO 6/131); fall negotiations are in Richards to Airy, 21 Oct. 1870; Airy to Richards, 22 and 28 Oct. 1870; Airy to Richards, 1 Nov. 1870; Richards to Airy, 8 Nov. 1870 (RGO 6/131); G. G. Stokes to Lockyer, 21 Nov. 1870 (JNL, eclipse Box 1, File "Eclipse 1870—Sicily"); Airy to Huggins, 12 Nov. 1870; Huggins to Ranyard?, 12 Nov. 1870 (RGO 6/131); Airy to Lockyer, 13 Mar. 1871 (RAS Papers, 51.1); application is Lockyer to Treasury, n.d. (RGO 6/131); quotation from Airy to Richards, 13 Nov. 1870 (RGO 6/131).

2. Quotation from Airy to Richards, 13 Nov. 1870; Airy describes his pessimism in Airy to Richards, 25 Nov. 1870 (RGO 6/131).

3. Sir Robert Stawell Ball, *In the high heavens* (London: Isbister and Co., 1901), 78.

4. The term "gentlemanly specialists" comes from Martin J. S. Rudwick, *The great Devonian controversy: The shaping of scientific knowledge among gentlemanly specialists* (Chicago: University of Chicago Press, 1985); the beliefs of that second generation are described in Frank M. Turner, "Public science in Britain, 1880–1919," *Isis* 71 (1980), 589–608.

5. One exception is Emmanuel Liais, a Frenchman whose career in Brazil began with observations of the eclipse of 1858: Lewis Pyenson, "Functionaries and seekers in Latin America: Missionary diffusion of the exact sciences, 1850–1930," *Quipu* 2 (1985), 393–96. The U.S. Naval Observatory's J. M. Gillis observed the eclipse of 1858 in Chile: copy of newspaper report on Gillis (RGO 6/115:162).

6. Eclipse of 1851, George Airy, "Suggestions to astronomers for the observation of the total eclipse of the sun on 1851 July 28," handwritten ms., 13 Dec. 1850; Airy to H. M. Addington, 5 June 1851 (RGO 6/119); R. C. Carrington, *An account of the late total eclipse of the sun on July 28, 1851* (Durham: Andrews, 1851); H. Beaufort to Airy, 28 Aug. 1851 (RGO 6/119); eclipse of 1858, Airy to Challis, 2 Apr. 1858; Airy to Foster Nash, 2 Apr. 1858; Airy to Capt. Washington, 2 Apr. 1858; Nash to Airy, 3 Apr. 1858; Nash to Airy, 16 Apr. 1858; Washington to Airy, 27 and 28 Apr. 1858 (RGO 6/115); Carrington to Airy, 27 Apr. 1858; Airy to Carrington, 15 May 1858; Carrington to Airy, 19 May 1858 (RGO 6/115); Carrington, *Information and suggestions addressed to persons who may be able to place themselves within the shadow of the total eclipse of the sun of September 7, 1858* (London: Eyre and Spottiswoode, 1858).

7. Airy's later alliance with the Society for Opposing the Endowment of Research, and his opposition of public funding for "investigations of undefined character, and at best of doubtful utility," is described in Roy MacLeod, "The support of Victorian science: The endowment of research movement in Great Britain, 1868–1900," *Minerva* 4 (1971), 197–230, quotation on 224–25.

8. Navy and customs, Airy to Lord John Russell, 22 Feb. 1860; Russell to Airy, 21 Mar. 1860, 9 Apr. 1860; Airy to Lord Clarence Paget, 1 May 1860;

Airy to John C. Adams, 30 May 1860; Admiral Washington to Airy, 12 June 1860 (RGO 6/123); communications with observers, Airy circular, 11 May 1860, 11 June 1860 (RGO 6/123); Airy to Charles Vignoles, 9 Feb. 1860; quotation from Airy to Joseph Beck, 12 May 1860 (RGO 6/123).

9. David Mackay, *In the wake of Cook: Exploration, science and empire, 1780–1801* (London: Croom Helm, 1985); Robert Stafford, *A scientist of empire* (Cambridge: Cambridge University Press, 1989).

10. Tennant's plans, J. Tennant to Airy, 1 Feb. 1867; Airy to Tennant, 12 Feb. 1867; Andrew Scott Waugh to Airy, 1 Feb. 1867; Airy to Tennant, 9 Mar. 1867; Tennant to Airy, 11 Mar. 1867; Airy to Tennant, 26 Mar. 1867; Browning to Airy, 10 Apr. 1867; Airy to Browning, 12 Apr. 1867, 21 May 1867; Airy to De la Rue, 1 Apr. 1867; De la Rue to Airy, 2 and 5 Apr. 1867, 13 Aug. 1867; Huggins to Airy, 15 Apr. 1867; Tennant to Airy, 4, 13, and 23 Apr. 1867, 11 Sept. 1867; Airy to Tennant, 24 Apr. 1867 (RGO 6/122); Airy's advocacy, Airy to undersecretary for India, 15 June 1867; Airy to Tennant, 15 June 1867; Tennant to Airy, 15 June 1867; India Office to Airy, 12 July 1867; Tennant to Airy, 15 July 1867; Tennant to undersecretary for India, 23 July 1867 (RGO 6/122); funding, India Office to Airy, 12 July, 21 Sept., 12 Oct. 1867; War Office to India Office, 10 Aug. 1867; Airy to India Office, 23 Sept. and 12 Oct. 1867; Airy to Tennant, 13 July 1867 (RGO 6/122).

11. *Synopsis of the contents of the British Museum*, 47th ed. (London: G. Woodfall and Son, 1844), 3–39, quotations on 4, 10; Edward Miller, *That noble cabinet: A history of the British Museum* (London: Andre Deutsch, 1973), chap. 9; attendance figures, "Return of the number of persons admitted to visit the British Museum," in *House of Commons parliamentary papers*, 1863 (455) xxix: 211, 1874 (157) li: 857, 1875 (416) lix: 679.

12. The literature on professions and professionalization is immense. On the sociological side, Philip Elliott, *The sociology of the professions* (New York: Herder and Herder, 1972) is useful. For the history of professions in Britain, see William Joseph Reader, *Professional men: The rise of the professional classes in nineteenth century England* (New York: Basic, 1966); Harold Perkin, *The rise of professional society: England since 1880* (London: Routledge, 1989).

13. "Address delivered by the President [E. J. Stone] on presenting the Gold Medal of the Society to Mr. Common," *RASMN* 44 (1884), 221–23, quotations on 223 and 222; laissez-faire science, Frank Turner, "Public science in Britain, 1880–1919," *Isis* 71 (1980), 589–608, on 591; amateur science, John Lankford, "Amateurs and astrophysics: A neglected aspect in the development of a scientific specialty," *SSS* 11 (1981), 275–303. "Gentlemanly specialists" likewise dominated geology until the end of the nineteenth century. For them, amateur status was not a hindrance, but a help: their prominence was due in part to a subculture that valued hard fieldwork and promoted intense competition over credit and recognition for new discoveries. Roy Porter, "Gentlemen and geology: The emergence of a scientific career, 1660–1920," *Historical journal* 21 (1978), 809–36; Rudwick, *The great Devonian controversy*, esp. chap. 2.

14. Andrew Thomas Gage and William Thomas Stearn, *A bicentenary history of the Linnean Society of London* (London: Linnean Society, 1988); Martin Rudwick, "The foundation of the Geological Society of London: Its scheme for cooperative research and its struggle for independence," *BJHS* 1 (1963), 325–55; Dreyer and Turner, eds., *History of the Royal Astronomical Society*. Many founders were dissatisfied with the Royal Society's treatment of their fields. The dissatisfactions could also be social: the London Electrical Society was founded in 1837 as a society for electricians, demonstrators, and instrument-builders (the manufacturing and artisan classes of science) in conscious opposition to the polite style of electrical research and discourse practiced in the Royal Society. Iwan Morus, "Currents from the underworld: Electricity and the technology of display in early Victorian England," *Isis* 84 (1993), 50–69.

15. Arnold Thackray, "Natural knowledge in cultural context: The Manchester model," *AHR* 79 (1974), 672–709; Ian Inkster, "Variations on a theme by Thackray," *British Society for the History of Science newsletter* 8 (1982), 15–18; on Edinburgh, Steven Shapin, "'Nibbling at the teats of science': Edinburgh and the diffusion of science in the 1830s," in Ian Inkster and Jack Morrell, eds., *Metropolis and province: Science in British culture, 1780–1850* (University of Pennsylvania Press, 1983), 151–78; on science and conservative provincials, Michael Neve, "Science in a commercial city: Bristol, 1820–1860," in Inkster and Morrell, eds., *Metropolis and province*, 179–204.

16. David Elliston Allen, *The naturalist in Britain: A social history* (London: Allen Lane, 1976); Lynn Merril, *The romance of Victorian natural history* (Oxford: Oxford University Press, 1989); John R. R. Christie, "Ideology and representation in eighteenth-century natural history," *Oxford art journal* 13 (1990), 3–10.

17. For example, the Linnean Society was organized at the Marlborough Coffee House, the Society for the Promotion of Natural History at the York Coffee House, the Society of the Arts at Rawthmell's Coffeehouse, and the Geological Society and Astronomical Society were both founded at Freemason's Tavern: Gage and Stearn, *A bicentenary history of the Linnean Society of London*, 4, 7; Bernard H. Becker, *Scientific London* (New York: D. Appleton, 1875), 54, also 89, 240, 315; Rudwick, "The foundation of the Geological Society of London," 328; Dreyer and Turner, eds., *History of the Royal Astronomical Society*, 35. On other clubs, see N. G. Coley, "The animal chemistry club: Assistant society to the Royal Society," *Notes and records of the Royal Society of London* 22 (1967), 173–85. To modern readers, a tavern might seem an unlikely place to launch a scientific society, but many nineteenth-century London debating and discussion societies, and poorer and less prestigious scientific clubs, were run in taverns: see Mark Girouard, *Victorian pubs* (London: Studio Vista, 1975), 12.

18. Becker, *Scientific London*, quotations on 24, 25, 26. For another example of the social life of societies, see Allen, *The naturalist in Britain*, 173–74.

19. William J. Ashworth, "The calculating eye: Baily, Herschel, Babbage and the business of astronomy," *BJHS* 27 (1994), 409–41; Jack Morrell and

Arnold Thackray, *Gentlemen of science: Early years of the British Association for the Advancement of Science* (Oxford: Clarendon Press, 1981). On the history of early scientific societies, see also David Philip Miller, "The social history of British science: After the harvest?" *SSS* 14 (1985), 115–35.

20. Morris Berman, *Social change and scientific organization: The Royal Institution, 1799–1844* (Ithaca: Cornell University Press, 1978); D. S. L. Cardwell, *The organization of science in England: A retrospect* (London: William Heinemann, 1957), chap. 3. (Despite their prominent place in the historiography of Victorian science, working-class institutes had their contemporary critics who lamented their lack of organization, systematic teaching, and facilities: see, for example, Becker, *Scientific London*, chap. 6 passim.) Anne Secord, "Corresponding interests: Artisans and gentlemen in 19th-century natural history," *BJHS* 27 (1994), 383–408; Secord, "Science in the pubs: Artisan botanists in early 19th century Lancashire," *History of science* 32 (1994), 269–315.

21. J. N. Hays, "The London lecturing empire, 1800–1850," in Inkster and Morrell, eds., *Metropolis and province*, 91–119; Ian Inkster, "Science and society in the metropolis: A preliminary examination of the social and institutional context of the Askesian Society of London, 1796–1807," *Annals of science* 34 (1977), 1–32, esp. 5–13; museums and cabinets, Richard D. Altick, *The shows of London* (Cambridge: Belknap Press, 1978), chaps. 25–34; London's technical marvels, see Andrew Saint, "The building arts of the first industrial metropolis," in Celina Fox, ed., *London: World city 1800–1840* (New Haven: Yale University Press, 1992), 51–76. By 1839, these lectures and museums were deemed important enough to attract calls to "extend opportunities of 'mental recreation and development among the people'" by lowering or eliminating their fees: "Report of the Committee of the Society for obtaining free Admission to National Monuments, and public Edifices containing works of Art," *The Gardener's magazine and register of rural and domestic improvement* 15 (July 1839), 417. However, some critics claimed that audiences were mainly interested in the commercial potential of science, or in appearing fashionable: one argued that while "forty years ago, in the days of Sir Humphrey Davy's triumphs, there was doubtless a great deal of scientific affectation in the fashionable world," in 1851 "every district of the town must have its scientific institute, and every suburb its courses of lectures": William Johnston, *England as it is: Political, social and industrial in the middle of the nineteenth century* (London: John Murray, 1851), 247.

22. Susan Sheets-Pyenson, "A measure of success: The publication of natural history journals in early Victorian Britain," *Publishing history* 9 (1981), 21–36; Sheets-Pyenson, "Popular scientific periodicals in Paris and London: The emergence of a low scientific culture, 1820–1875," *Annals of science* 42 (1985), 549–72.

23. Charles Darwin was exposed to materialism and other intellectual contraband through the radical, but intellectually serious, student underground at Edinburgh: see Adrian Desmond and Jim Moore, *Darwin* (London: Michael Joseph, 1991), chap. 4.

24. Radical biology is discussed in Adrian Desmond, "Artisan resistance and evolution in Britain, 1819–1848," *Osiris* (2d series) 3 (1987), 77–110; Desmond, *The politics of evolution: Morphology, medicine, and reform in radical London* (Chicago: University of Chicago Press, 1989); on astronomy, see Simon Schaffer, "The nebular hypothesis and the science of progress," in Jim Moore, ed., *History, humanity and evolution* (Cambridge: Cambridge University Press, 1989), 131–61; Schaffer, "The leviathan of Parsontown: Literary technology and scientific representation," in Tom Lenoir, ed., *Inscribing science: Scientific texts and the materiality of communication* (Stanford: Stanford University Press, 1998), 182–222.

25. John Laurent, particularly his "Science, society, and politics in late 19th-century England: A further look at the Mechanics' Institutes," *SSS* 14 (1984), 585–619. Laurent has also found similar patterns in Australian working-class institutes: see "Bourgeois expectations and working class realities: Science and politics in Sydney's schools of art," *Journal Royal Australian Historical Society* 75 (1989), 33–50.

26. The definitive biography of Lockyer is A. J. Meadows, *Science and controversy: A biography of Sir Norman Lockyer* (Cambridge: MIT Press, 1972).

27. Brett to Ranyard, 18 Nov. 1870 (RAS Papers 51.1); Stokes to Airy, 20 Nov. 1870 (RGO 6/131).

28. Pritchard to Lockyer, 16 Nov. 1870 (RAS Papers 51.2); Pritchard to Airy, 16 Nov. 1870 (RGO 6/131); Clifford to Lockyer, n.d.; W. H. Hudson to Ranyard, 17 Nov. 1870 (RAS Papers 51.2); John Tyndall, *Hours of exercise in the Alps* (New York: D. Appelton, 1897), chap. 8. There was some coordination between the groups: see Ranyard to Airy, 14, 21, 24, 30 Nov. 1870, 2 Dec. 1870 (RGO 6/131). Father Steven Perry, S.J., director of the Stonyhurst Observatory, put off a magnetic survey of Belgium for a year to participate: Perry to Lockyer, 13 July 1870 (JNL, Box "Lockyer Letters," File "P").

29. Airy to Lockyer, 3 Dec. 1870 (JNL, Box "Lockyer Letters," File "A"); "The eclipse expedition," *Nature* (8 Dec. 1870), 101.

30. Airy to Meriwale, 31 May 1871 (RGO 6/134); Edwin Durkin to Lockyer, 29 June 1871 (JNL, eclipse Box 1, File "Eclipse 1871—Packet 1"); Lockyer to Airy, 5, 11, 17 July 1871 (RGO 6/134).

31. *Nature* (15 June 1871), 128; A. J. Meadows, *Science and controversy*, 68–69; Lockyer, draft of Treasury proposal, n.d. (JNL, eclipse Box 1, File "Eclipse 1871—Packet 1"). See also Minutes of Solar Eclipse Committee meeting, 20 July 1871 (RS CMB.2, "Miscellaneous Committees, 1869–1884"); Stokes to Lockyer, 9 July 1871 (JNL, eclipse Box 1, File "Eclipse 1871—Packet 1").

32. Stokes to Airy, 25 July 1871; Airy to Stokes, 27 July 1871; Airy to Lassell, 31 July 1871; identical to Stokes, 31 July 1871 (RGO 6/134); Stokes to Lockyer, 28 July 1871 (JNL, eclipse Box 1, File "Eclipse 1871—Packet 1"); Lockyer to Airy, 9 Sept. 1871 (RGO 6/134); quotation from Lockyer to Young, 12 Sept. 1871 (CAY, Box 7, File 18, "Lockyer, J. Norman"). It is probably no coincidence that the British Association for the Advancement of Science presi-

dent, Sir William Thomson (later Lord Kelvin), included in his address at the annual meeting in Edinburgh a discussion of the advances made by eclipse observation in science's understanding of the sun: "Inaugural address of Sir William Thomson," *Nature* (3 Aug. 1871), 262–70; "The British Association meeting at Edinburgh," *Nature* (10 Aug. 1871), 290.

33. Eclipse of 1875, Stokes to Secretary of State for India, in Minutes of Eclipse Committee, 14 Jan. 1875, 15 Jan. 1875, 20 Jan. 1875, 28 Jan. 1875, 1 Feb. 1875, 4 Feb. 1875 (CMB.2, "Miscellaneous Committees 1869–1884"); Meadows, *Science and controversy*, 105–6.

34. A. J. Meadows, *Greenwich Observatory*, vol. 2: *Recent history (1836–1975)* (London: Taylor and Francis, 1975), 90. On expeditions in the 1700s, see Harry Woolf, *The Transits of Venus: A study of eighteenth-century science* (Princeton: Princeton University Press, 1959).

35. On Lockyer and the RAS in this period, see Meadows, *Science and controversy*, 108–12; Roy MacLeod, "The support of Victorian science: The endowment of research movement in Great Britain, 1868–1900," *Minerva* 4 (1971), 197–230.

36. The following analysis is inspired by the masterful statistical and sociological analysis of the Salem witchcraft trials in Paul Boyer and Stephen Nissenbaum, *Salem possessed: The social origins of witchcraft* (Cambridge: Harvard University Press, 1974).

37. Lockyer, "Instructions for observers" (JNL, eclipse Box 1, File "Eclipse 1871—Packet 1").

38. For example, in the Sabine papers at the Royal Society, there are files of correspondence with Stokes, De la Rue, and Huggins from the late 1760s and early 1870s, but these letters do not mention eclipses or committee business.

39. On fatigue, see Airy to Pritchard, 23 June 1870 (RGO 6/131.1); Airy to De la Rue, 5 Aug. 1871 (RGO 6/134); Airy to Lockyer, 7 Sept. 1871 (JNL, Box "Lockyer Papers," File "A"); loss of face, Airy to Richards, 13 Nov. 1870 (RGO 6/131); concern about government interest, Airy to Lockyer, 7 and 12 July 1871 (RGO 6/134). Lockyer also speculated that Airy was concerned to protect government support for transit of Venus expeditions, and eclipse fieldwork would advance at the transit's expense: "The approaching eclipse," *Nature* (3 Aug. 1871), 259.

40. Charles Pritchard to Airy, 23 June 1870 (RGO 6/131).

41. Technically this was a rump body, but, as Airy put it, "an illegal body which works is better than a legitimate body which will not work": Airy to Lockyer, 13 Mar. 1871 (RAS Papers, 51.1). See also G. G. Stokes to Lockyer, 21 Nov. 1870 (JNL, eclipse Box 1, File "Eclipse 1870—Sicily"); Airy to Huggins, 12 Nov. 1870 (RGO 6/131); Huggins to Ranyard?, 12 Nov. 1870 (RAS Papers, 51.1).

42. Davis, "Addenda to notes on eclipse photography," handwritten ms., n.d.; Brett, "Some particulars to be especially noted by those observers who make drawings of the corona," handwritten ms., n.d. (JNL, eclipse Box 1, File "Eclipse 1871—Packet 2").

43. On Huxley's position in the 1850s, see Adrian Desmond, *Archetypes and ancestors: Paleontology in Victorian London, 1850–1870* (Chicago: University of Chicago Press, 1982), esp. chap. 1; Lockyer, quoted in D. S. L. Cardwell, *The organization of science in England: A retrospect* (London: William Heinemann, 1957), 120. We should take Lockyer's words with a few grains of salt, since he was speaking before a commission investigating the state of science in Britain, and had a special gift of exaggeration.

44. Marie Boas Hall, *All scientists now: The Royal Society in the nineteenth century* (Cambridge: Cambridge University Press, 1984), chap. 3; Roy MacLeod, "The Royal Society and the Government Grant: Notes on the administration of scientific research, 1849–1914," *Historical journal* 14 (1971), 323–58. The German plan in particular served as a model for science in the United States as well: see, for example, Owen Hannaway, "The German model of chemical education in America: Ira Remsen at Johns Hopkins (1876–1913)," *Ambix* 23 (1976), 145–64.

45. Dreyer and Turner, *History of the Royal Astronomical Society*, 167–78; Meadows, *Science and controversy*, 92–103; Roy MacLeod, "The X-Club: A social network of science in late-Victorian England," *Notes and records of the Royal Society of London* 24 (1970), 305–22; J. Vernon Jensen, "The X Club: Fraternity of Victorian scientists," *BJHS* 5 (1970), 63–72; Ruth Barton, "The X Club: Science, religion and social change in Victorian England" (Ph.D. dissertation, History and Sociology of Science, University of Pennsylvania, 1986); Barton, "'An influential set of chaps': The X-Club and Royal Society politics, 1864–1885," *BJHS* 23 (1990), 53–81.

46. Romualdus Sviedrys, "The rise of physics laboratories in Britain," *HSPS* 7 (1976), 405–36; Sviedrys, "The rise of physical science at Victorian Cambridge," *HSPS* 2 (1970), 127–45, and comment by Arnold Thackray, *HSPS* 2 (1970), 145–49; science at Oxford, Janet Howarth, "Science education in late-Victorian Oxford: A curious case of failure?" *English historical review* 102 (1987), 334–71; F. G. Hopkins and Robert Kohler, *From medical chemistry to biochemistry: The making of a biomedical discipline* (Cambridge: Cambridge University Press, 1982), 49–55, quotation on 53.

47. Frank Turner, "Public science in Britain, 1880–1920," *Isis* 71 (1980), 589–608, quotation on 592.

48. This account summarizes Donald Mackenzie, "Eugenics in Britain," *SSS* 6 (1976), 499–532, quotations on 501. On Pearson, see Bernard Norton, "Karl Pearson and statistics: The social origins of scientific innovation," *SSS* 8 (1978), 3–34. See also Daniel Kevles, *In the name of eugenics* (Berkeley: University of California Press, 1985).

49. Frank M. Turner, "The Victorian conflict between science and religion: A professional dimension," *Isis* 69 (1978), 356–76, quotation on 370.

50. John Lankford, "Amateur versus professional: The transatlantic debate over the measurement of Jovian longitude," *BAA journal* 89 (1979), 574–82; Lankford, "Amateurs versus professionals: The controversy over telescope size in late Victorian science," *Isis* 72 (1981), 11–28; Norris Hetherington, "Ama-

teur versus professional: The British Astronomical Association and the controversy over canals on Mars," *BAA journal* 86 (1976), 303–8; Barton, "'An influential set of chaps,'" esp. 57–58, 69, 72–73; Barton, "'Huxley, Lubbock, and half a dozen others': Professionals and gentlemen in the formation of the X Club, 1851–1864," *Isis* 89 (1998), 410–44.

51. These figures are based on a count of obituaries in the *Monthly notices* from 1845 to 1868, and a tabulation of obituaries listed in the *General index to volumes XXX to LII of the Monthly notices of the Royal Astronomical Society, 1869–1892* (London: Royal Astronomical Society, 1896), 112–17. Between 1845 and 1868, 212 members died; of those, 32 (15.1 percent) were clergy (overwhelmingly Anglican, with a few Presbyterian and Catholic), and 45 (21.2 percent) were military officers. Between 1869 and 1892, 204 members died; of those, 24 (11.8 percent) were clergy, and 21 (10.3 percent) were army or navy officers. It is likely that clergy are underreported in these documents, since some were listed with the honorific of "Dr." rather than "Rev." The changes were more dramatic in the Royal Society and British Association: see Turner, "The Victorian conflict between science and religion," 366–67.

52. The British Astronomical Association's leaders—E. Walter Maunder, Nathaniel Green, William Huggins, Agnes Clerke, William Wesley, and others— were almost all FRAS, as were thirty-nine of the BAA's first fifty provisional members. F. J. Sellers and P. Doig, eds., "The history of the British Astronomical Association: The first fifty years," *BAA memoirs* 36 (1948), chap. 1; "Circulars issued by the Provisional Committee, II," *BAA journal* 1 (1890), 17, 19.

53. Desmond Bowen, *The idea of the Victorian church: A study of the Church of England, 1883–1889* (Montreal: McGill University Press, 1968); Alan Haig, *The Victorian clergy* (London: Croom Helm, 1984), esp. chap. 1. See also Frank M. Turner, "The Victorian conflict between science and religion," 368.

54. This summary draws on T. W. Heyck, "From men of letters to intellectuals: The transformation of intellectual life in nineteenth-century England," *Journal of British studies* 20 (1981), 158–83; Heyck, *The transformation of intellectual life in Victorian England* (Lyceum, 1982); John Gross, *The rise and fall of the man of letters* (New York: Macmillan, 1969).

55. On the controversy over state funding of astrophysical research and the Proctor-Lockyer affair, see Meadows, *Science and controversy*, 95–103; Dreyer and Turner, eds., *History of the Royal Astronomical Society*, 172–78.

56. On the BAAS and Devonshire Committees, see Meadows, *Science and controversy*, chap. 4.

57. J. J. Thompson, *Recollections and reflections* (London: G. Bell and Sons, 1936), 24–25.

58. For example, C. H. Stigand's classic *Hunting the elephant in Africa* (1913: repr. New York: St. Martin's, 1986) included discussions of big-game behavior, tracking abilities of various African tribes, and chapters on "Curious African sayings and ideas" and "Mimicry and protective colouration in insects." In the foreword, Theodore Roosevelt described Stigand as "a field natu-

ralist of unusual powers" (v). See also John Mackenzie, *Empire of nature: Hunting, conservation and British imperialism* (Manchester: Manchester University Press, 1988). For engineers, see Edward Sandes, *The military engineer in India*, 2 vols. (Chatham: Institution of Royal Engineers, 1935).

59. Susan Sheets-Pyenson, "Cathedrals of science: The development of colonial natural history museums during the late nineteenth century," *History of science* 25 (1986), 279–300; T. J. Barron, "Science and the nineteenth-century Ceylon coffee planters," *Journal of imperial and commonwealth history* 16 (1987), 5–21; medical services, Daniel Headrick, *The tentacles of progress: Technology transfer in the age of imperialism, 1850–1940* (Oxford: Oxford University Press, 1988), chap. 7; W. H. R. Rivers, "Report on anthropological research outside America," in *Reports upon the present condition and future need of the science of anthropology* (Washington: Carnegie Institution of Washington, Publication no. 200), 5–28, quotation on 7.

60. Robert Stafford, "Geological surveys, mineral discoveries, and British expansion, 1835–71," *Journal of imperial and commonwealth history* 12 (1984), 5–32; Stafford, "Roderick Murchison and the structure of Africa: A geological prediction and its consequences for British expansion," *Annals of science* 45 (1988), 1–40; and Stafford, *A scientist of empire*.

61. Richard Grove, "Conserving Eden: The (European) East India Companies and their environmental policies on St. Helena, Mauritius and in western India, 1660 to 1854," *CSSH* 35 (1993), 318–51. Grove's description of the intellectual and institutional conditions on Mauritius that encouraged the development of ecological thinking is similar to the work on "local environments" in Robert E. Kohler, "Innovation in normal science: Bacterial physiology," *Isis* 76 (1985), 162–81.

62. Mark Harrison, "Tropical medicine in nineteenth-century India," *BJHS* 25 (1992), 299–318; Philip Curtin, "Medical knowledge and urban planning in tropical Africa," *AHR* 90 (1985), 594–613; John Cell, "Anglo-Indian medical theory and the origins of segregation in West Africa," *AHR* 90 (1985), 307–35; Michael Worboys, "The emergence of tropical medicine: A study in the establishment of a scientific specialty," in G. Lemaine, ed., *Perspectives on the emergence of scientific disciplines* (The Hague, 1976), 76–98. Despite the unquestionable power of Western medicine to deal with individual patients' emergencies and diseases, at the social and ecological level colonial medicine's record is mixed. In precolonial East Africa, a system of land use developed by local rulers that had contained tsetse flies and sleeping sickness broke down with European invasion, and sleeping sickness returned to human population centers. Colonial medical services later had to deal with problems caused by the disruption of native political systems and warfare. See Gwyn Prins, "But what was the disease? The present state of health and healing in African studies," *Past and present* 124 (1989), 159–79; John Ford, *The role of the trypanosomiases in African ecology: A study of the tsetse fly program* (Oxford: Oxford University Press, 1971).

63. S. N. Sen, *Scientific and technical education in India, 1781–1900* (New

Delhi: Indian National Science Academy, 1991), chap. 9; Edward W. Ellsworth, *Science and social science research in British India, 1780–1880: The role of Anglo-Indian associations and government* (New York: Greenwood Press, 1991), chaps. 6–8; T. J. Barron, "Science and the nineteenth-century Ceylon coffee planters," *Journal of imperial and commonwealth history* 16 (1987), 5–21.

64. Daniel Headrick, *The tools of empire: Technology and European imperialism in the nineteenth century* (Oxford: Oxford University Press, 1981); Headrick, *Tentacles of progress*, chap. 3. It should be noted that native Indian travel flourished despite a price structure that enforced European-Indian segregation and confined all but the wealthiest Indians to crowded third-class carriages.

65. Sandes, *The military engineer in India*, vol. 2, 231–33; Anon., "The Indian Trigonometric Survey," *Van Nostrand's engineering magazine* 13 (1875), 367–71, esp. 370–71; Mildred Archer, *Natural history drawings in the India Office Library* (London: Her Majesty's Stationery Office, 1962). This was part of a broader traffic between Western and Indian art in which "empirical naturalism" of European natural history figures largely. For a perceptive analysis of this traffic in religious portraiture, see Martina Barash, "Mixed company: Two portraits of Hindu holy men and syncretic painting in British India" (B.A. thesis, Harvard University, 1988). My thanks to Ms. Barash for calling this work to my attention.

66. On physicians, see Mark Harrison, *Public health in British India: Anglo-Indian preventive medicine, 1859–1914* (Cambridge: Cambridge University Press, 1994), 11–12; on Western physician-scientists in India, Philip Meadows Taylor, *Story of my life*, 2 vols. (London: 1877). On military advisors, see Geoffrey Parker, *The military revolution: Military innovation and the rise of the West, 1500–1800* (Cambridge: Cambridge University Press, 1988), 128–36; I. B. Watson, "Fortifications and the idea of force in early English East India Company relations with India," *Past and present* 88 (1980), 70–87; Gayle Ness and William Stahl, "Western imperialist armies in Asia," *CSSH* 19 (1977), 2–29; John Pemble, "Resources and techniques in the Second Maratha War," *Historical journal* 2 (1976), 375–404. No study yet exists that places non-Western patronage of science and technology in the context of ritual exchanges and patronage that defined court culture, but the excellent recent work on science and the early modern European court could serve as a fine model: see, in particular, Robert Westman, "The astronomer's role in the sixteenth century: A preliminary study," *History of science* 18 (1980), 105–47; Mario Biagoli, *Galileo, courtier: The practice of science in the culture of absolutism* (Chicago: University of Chicago Press, 1993); Pamela Smith, "Alchemy as a language of mediation at the Habsburg court," *Isis* 85 (1994), 1–25.

67. Satpal Sangwan, "Indian responses to European science and technology, 1757–1857," *BJHS* 21 (1988), 211–32; S. N. Sen, *Scientific and technical education in India, 1781–1900* (New Delhi: Indian National Science Academy, 1991), 304–68, quotation on 333, expenditures in table 8.1. The view that

British policies were designed to keep Indians from learning science is expressed in Sir Saiyid Ahmad to E. White, 22 Feb. 1885, in J. P. Naik, ed., *Selections from the educational records of the Government of India*, vol. 2: *Development of university education, 1860–1887* (Delhi: National Archives of India, 1963), 367–74. It is also notable that native elites were supporting science while British intellectuals were occupied with the European-Orientalist education debate: Sangwan, *Science technology and colonisation: An Indian experience, 1757–1857* (Delhi: Anamika Prakashan, 1991), chap. 3.

68. H. H. Turner to Michael Foster, 28 Feb. 1894, copied in Solar Eclipse Committee minutes, 2 May 1894 (RAS 54.1).

69. Solar Eclipse Committee minutes, 10 May 1889 (RAS 54.1); Turner, application to Government Grant Committee, 1 Feb. 1889; Evan Macgregor to E. B. Knobel, 12 Mar. 1889; Knobel to Maunder, 27 Apr. 1889; Knobel to Foreign Office, 13 Aug. 1889; A. A. Common to Knobel, 9 Sept. 1889 (RAS Papers 53, "Eclipse of 1889").

70. Solar Eclipse Committee minutes, 6 Apr. 1892, 5 May 1893 (RAS 54.1); A. A. Common, "Preliminary report of the Joint Solar Eclipse Committee," *RASMN* 53 (1893), 472–73.

71. H. H. Turner to Michael Foster, 28 Feb. 1894, copied in Solar Eclipse Committee minutes, 2 May 1894 (RAS 54.1); Foster to Turner, 20 Jan. 1894; Turner to Foster, 28 Feb. 1894; Foster to Turner, 19 Mar. 1894, copied in Solar Eclipse Committee minutes, 2 May 1894 (RAS 54.1); Foster to Turner, 19 Mar. 1894; Turner to Foster, 24 Apr. 1894, copied in Solar Eclipse Committee minutes, 2 May 1894 (RAS 54.1).

72. Foster to Turner, 19 Mar. 1894; Foster to Turner, 28 Apr. 1894, copied in Solar Eclipse Committee minutes, 2 May 1894 (RAS 54.1).

73. Hence, when in 1900 Norman Lockyer asked committee secretary Edmond H. Hills whether South Kensington or the Royal Society should ask the Foreign Office for landing rights in Spain, Hills replied, that "the Science and Art Department have, as far as I know, nothing to do with it"; Norman Lockyer to Edmond Hills, ? Feb. 1900; Hills to Lockyer, 19 Feb. 1900; acknowledgments, Hills to Royal Society, 5 June 1898 (RAS 55.2).

74. Turner to Foster, 28 Feb. 1894, copied in Solar Eclipse Committee minutes, 2 May 1894 (RAS 54.1).

75. This overlap was engineered to give expedition members a voice in planning: Hills to Junior Institute of Engineers, 29 Jan. 1905 (RAS Papers 55.2).

76. Turner to Lockyer, 14 Aug. 1890 (JNL, "Lockyer Letters," File T). Lockyer and Fowler also traveled together to visit Stonehenge in 1901: Meadows, *Science and controversy*, 251.

77. Christie to Hills, 19 Mar. 1897 (RGO 7/197).

78. RAS officers and Gold Medal winners are listed in Dreyer and Turner, *History of the Royal Astronomical Society*, 207, 250–53. Royal Society Government Grants Committee members are listed in "Government Grant Committee minutes" (RSL, CBM.66.c) and "Government Grant Committee min-

utes, Board B" (RSL, CBM.1336-1). On the history of the grant, see Roy Mac-Leod, "The Royal Society and the Government Grant," 323–58.

79. Lankford, "Amateurs versus professionals," 12–13.

80. Lankford, "Amateurs and astrophysics," 275–303; John E. Hodge, "Charles Dillon Perrine and the transformation of the Argentine National Observatory," *JHA* 8 (1977), 12–25; F. J. M. Stratton, "John Evershed," *Biographical memoirs of the fellows of the Royal Society* 3 (1957), 41–51.

81. J. Norman Lockyer, *Recent and coming eclipses: Being notes on the total solar eclipses of 1893, 1896, and 1898* (London: Macmillan and Co., 1897), 8.

82. My analysis of this work is influenced by John Law, "Technology and heterogeneous engineering: The case of Portuguese expansion," in Wiebe Bijker, Thomas Hughes, and Trevor Pinch, eds., *The social construction of technological systems: New directions in the sociology and history of technology* (Cambridge: MIT Press, 1987), 111–34.

83. Airy to H. Meriwale, 31 May 1871 (RGO 6/134).

84. Airy to Lord John Russell, 23 Feb. 1860 (RGO 6/123).

85. *Daily Mail* (London), 14 Nov. 1870 (RGO 6/132). The Americans, in their appeal to the U.S. government for funds to observe the eclipse of 1869, had played the same cards: Harvard professor Joseph Winlock argued that the "scientific world of Europe . . . would be watching and expecting great results" from American scientists. Winlock, quoted in Bessie Zaban Jones and Lyle Gifford Boyd, *The Harvard College Observatory: The first four directorships, 1839–1919* (Cambridge: Harvard University Press, 1971), 163.

86. On the National Academy of Sciences, see A. Hunter Dupree, *Science and the federal government* (1955; repr. University of Chicago Press, 1988); on American astronomical societies, Marc Rothenberg, "Organization and control: Professionals and amateurs in American astronomy, 1899–1918," *SSS* 11 (1981), 305–25. On the importance of local forces in shaping American science, see Robert Kohler, "The Ph.D. machine: Building on the collegiate base," *Isis* 81 (1990), 638–62; on the role of local environments in shaping research programs, see Kohler, "Innovation in normal science: Bacterial physiology," *Isis* 76 (1985), 162–81.

87. Airy to H. Meriwale, 31 May 1871 (RGO 6/134). On science and nineteenth-century justifications of empire, see Michael Adas, *Machines as the measure of men: Science, technology, and ideologies of Western dominance* (Ithaca: Cornell University Press, 1989), esp. chaps. 3–5.

88. "Path of the total phase of the solar eclipse, August 17–18, 1868, from Aden to the Torres Straits," Nautical Almanac Circular 11, 21 Oct. 1867 (RGO 6/122).

89. Tennant, three maps of India with path of 1868 eclipse drawn in and other annotations, n.d. (ca. Jan. 1867) (RGO 6/122).

90. Map of India, Society for the Diffusion of Useful Knowledge, n.d. (RGO 6/134).

91. De la Rue to Airy, 11 Oct. 1861 (RGO 6/123).

92. Draft of letter to Treasury, n.d., ca. 1871 (JNL, eclipse Box 1, File "Eclipse 1871—Packet 1").

93. Ball, *In the High Heavens*, 84.

94. John Tennant to J. Norman Lockyer, 29 Nov. 1871 (JNL, File "Eclipse 1871—Packet 1"); War Ministry to Lockyer, 31 Mar. 1882 (JNL, File "Eclipse 1882—Egypt").

95. James Tennant to J. Norman Lockyer, 29 Nov. 1871 (JNL, File "Eclipse 1871—Packet 1"); War Minister to Lockyer, 31 Mar. 1882 (JNL, File "Eclipse 1882–Egypt"); E. C. Pickering to Grove-Hills, 29 Dec. 1898 (RAS Papers 55.2); Meteorological Office, Simla, to Corbett Gore, 23 July 1900 (RAS Papers 55.2).

96. Foster Nash to Airy, 3 Apr. 1858 (RGO 6/121); Adm. Ommaney to William Lassell, 3 Dec. 1870 (RAS Papers 51.2); Madras government to R. Abbay, 21 Nov. 1871 (JNL, File "Eclipse 1871—Packet 1"); Earl Granville to J. Norman Lockyer, 23 May 1882 (JNL, File "Eclipse 1882–Egypt"); A. H. Lugard to Ralph Copeland, 25 Sept. 1897; L. A. Lugard to A. H. Lugard, 30 Sept. 1897 (ROE, 36.262); James Keeler to Edmond Grove-Hills, 16 Dec. 1898 (RAS Papers 55.2). Army officers, Tennant to Lockyer, 1 Nov. 1871 (JNL, eclipse Box 1, File "Eclipse 1871—Packet 1"); Charles Strahan to Christie, 17 Feb. 1897; S. G. Burrard to F. W. Dyson, 24 June 1896, 14 Sept. 1897, 23 Sept. 1897 (RGO 7/195); L. A. Lugard to Ralph Copeland, 25 Sept. 1897; L. A. Lugard to A. H. Lugard, 30 Sept. 1897; L. A. Lugard to Sir Charles Lyall, 25 Nov. 1897 (ROE, 36.262, "Ralph Copeland Correspondence 1896–1897"); railroad engineers, Charles Vignoles to Airy, 8 Feb. 1860, 24 Mar. 1860 (RGO 6/124); K. H. Vignoles, *Charles Blacker Vignoles: Romantic Engineer* (Cambridge: Cambridge University Press, 1982), passim; W. N. Shaw to Campbell, 21 Mar. 1906 (MLS, Box "Flint Island #1," File 1).

97. James Tennant to Lockyer, 1 Nov. 1871 (JNL, eclipse Box 1, File "Eclipse 1871—Packet 1"). On work of local informants and rumor-control, see Campbell, "In the shadow of the moon" (MLS, typescript m.s., n.d.), 65–84, 96–97, 107, 110, 118; Charles Young, "An astronomer's summer trip," *Scribner's magazine* (July 1888), 90–95. Quotation is from Lockyer, "The total eclipse of the sun, April 16th, 1893," *Philosophical transactions*, series A, 187 (1896), 567. They also played this role for American expeditions: see Pang, "Culture, Gender, and Astrophysical Fieldwork," esp. 30–37.

98. Petition, "The Memorial of the Undersigned Fellows of the Royal Society and Royal Astronomical Society, and Students of Science," n.d.; Lockyer to vice-admiral, C-in-C Malta, 19 Dec. 1870 (JNL, "Eclipse 1870—Sicily"). Other expressions of personal admiration are in Young, "An astronomer's summer trip," 93; Campbell, "In the shadow of the moon," 125–26.

99. Airy to De la Rue, 5 Aug. 1871 (RGO 6/134).

100. Pogson even sent his son to England to study astrophotography, and both De la Rue and Airy quickly made themselves unavailable to see him: Pogson to Airy, 29 Mar. 1871; Airy to Pogson, 27 May 1871; Pogson to Airy, 1 July 1871; Pogson to De la Rue, 1 July 1871; De la Rue to Pogson, 2 Aug. 1871; Norman Pogson Jr. to Airy, 16 Oct. 1871 (RGO 6/134).

101. Henry Roscoe, recommendation letter for Arthur Schuster, 8 Nov. 1879 (RS, Arthur Schuster papers).

102. Quotations from Joseph Beck to Airy, 5 May 1860; Airy to Beck, 12 May 1860 (RGO 6/123). Airy to Rev. H. A. Atwood, 23 June 1860; Airy to John Breen, 25 May 1860 (RGO 6/124). Wives of volunteers were excluded unless they could serve as observers or assistants: see Francis Galton to Airy, 14 June 1860; Airy to Galton, 16 June 1860 (RGO 6/124).

103. Quotation from Airy to Warren De la Rue, 19 Mar. 1867; Airy to Huggins, 19 Mar. 1867 (RGO 6/122); Huggins to Airy, 23 Mar. 1867; De la Rue to Airy, 30 Mar. 1867; John Browning to De la Rue, 28 Mar. 1867 (RGO 6/122).

104. Airy to Lassell, 20 May 1870 (RGO 6/131).

105. E. H. Hills to Ralph Copeland, 22 Feb. 1900 (ROE, 36.264 "Ralph A. Copeland Correspondence—Letters in 1900").

106. G. H. Richards to Airy, 7 July 1870 (RGO 6/131); Huggins to Airy, 9 July 1870 (RGO 6/131); Airy to Richards, 12 July 1870 (RGO 131.5). This was the source of a dispute between the JPEC and Scottish Astronomer Royal Ralph Copeland, when Copeland took his mechanic to India: Hills to Copeland, 4 May 1898, 22 Apr. 1900 (ROE 36.263, "Ralph Copeland Correspondence 1898–99"). Copeland explained that his instruments "were not complete machines such as are turned out by our best opticians, but rather combinations of laboratory and observatory apparatus put together for use on this particular occasion. This made it all the more desirable that the person who was to assist in operating them should be perfectly familiar with all their details." Copeland to Hills, 14 May 1898 (ROE 34.257, "Letterbook 1896–1911"). This echoes appeals to tacit knowledge explored by Harry Collins in *Changing order: Replication and induction in scientific practice* (Beverly Hills: Sage, 1985), esp. chap. 3.

107. Henry Davis, "Addenda to 'Notes on eclipse photography,'" handwritten ms., n.d., ca. 1871 (JNL, Eclipses Box 1, File "Eclipse 1871—Packet 2").

108. *Nature* opined that "naval officers are so familiar with optical instruments, that they need but few hints to make most of the observations required during eclipses": "The total eclipse of the sun," *Nature* (17 Feb. 1898), 366.

109. Quotation from Lockyer to Hills, Feb. 1900 (RAS Papers, 55.2); see also Lockyer, "What the 'Melpomenes' did at Viziadrug," *Journal Royal United Service Institution* (15 Aug. 1899), 861–77; Lockyer to Copeland, 13 Mar. 1900 (ROE, 36.264 "RAC Correspondence—Letters in 1900").

110. P&O Company to Lockyer, 25 Mar. 1882 (JNL, "Eclipses Box 1," File "Eclipse 1882—Egypt"); Turner to Elder Dempster and Co., 20 Sept. 1889; Turner to Royal Mail Steam Company, 23 Sept. 1889 (RAS Papers, 53.4).

111. R. Wilby to Lockyer, 19 Nov. 1870 (JNL, "Eclipses Box 1," File "Eclipse 1870—File 1").

112. W. W. Campbell, "The Crocker eclipse expedition of 1908," *Lick Observatory bulletin* 131 (1909), 1–5.

113. Texas and Pacific Railway pass, n.d. (DPT, Box 26, "Eclipse of 29 July 1878," File "Legal-Personal"); Lewis Tuttle to David Todd, 21 May 1887 (DPT, Box 30, "Eclipse of 1887," File 194, "Correspondence 12/4/86–6/30/87").

114. Burrard to Dyson, 24 June 1896 (RGO 7/195); Map of "Savannah Line," n.d. (MLS, Box 5, "Eclipse—Thomason Ga., 28 May 1900"); "Oceanic Steamship Company's round the world map," n.d. (MLS, Box "Flint Island #1," File 1B).

115. Huggins to Airy, 9 July 1870 (RGO 6/131); Airy to Richards, 12 July 1870 (RGO 131.5).

116. G. H. Richards to Airy, 7 July 1870 (RGO 6/131).

117. Hills to Copeland, 22 Apr. 1900 (ROE, 36.264 "Ralph A. Copeland Correspondence—Letters in 1900").

118. Warren De la Rue to George Airy, 4 July 1860 (RGO 6/123).

119. "Eclipse expedition contents of boxes," n.d. (MLS, Box 15, File "Eclipse-Chile 1893").

120. P&O Company to Copeland, 27 Nov. 1897 (ROE, 36.262 "Ralph Copeland Correspondence 1896–1897").

121. See "Lady Astronomer" (Miss Brown), *In pursuit of a shadow* (London: Trübner and Co., 1891), chap. 1; assurances given by J. W. Riddle to Campbell, 18 May 1905 (MLS, Box 1, "Spanish eclipse," File "Letters of Preparation 1904 and 1905").

122. J. P. Maclear to J. Norman Lockyer, 24 Dec. 1871 (JNL "Eclipse 1871— Packet 1"). One Dutch astronomer even asked James Keeler if he could "attend the eclipse of 1900 . . . to study your American instruments, and your skillful and practical methods of construction, packing, and mounting." Albert Nijland to James Keeler, 22 Feb. 1900 (MLS, Box 43, File "Albert Nijland").

123. George Ellis to David Todd, 28 Apr. 1887 (DPT, Box 30, "Eclipse of 1887," File "Legal-Personal"); Campbell to F. R. Zeil, n.d. (MLS, Box "Flint Island #1," File 1C).

124. J. Glasenapp (St. Petersburg Observatory) to David Todd, 4 Dec. 1886; R. Kukio to Todd, 26 May 1887 (DPT, Box 30, "Eclipse of 1887," File 194, "Correspondence 12/4/86–6/30/87"); Foreign Office to Knobel, 11 May 1889 (RAS Papers 53.2); A. Nicholson to Marquis of Landsdowne, 12 May 1905 (RAS Papers 55.2); Instituto de Observatoria, San Fernando to Campbell, 22 May 1905 (MLS, Box 1, "Spanish eclipse," File "Letters of Preparation 1904 and 1905").

125. G. A. Leland to David Todd, 27 Apr. 1887 (DPT, Box 30, "Eclipse of 1887," File "Correspondence 12/4/86–6/30/87").

126. "Contents of trunk," n.d. (JNL, eclipse Box 2, File "Eclipse 1886").

127. Newcomb to Lockyer, 15 July 1878 (SNP, Box 4, "Letterbooks, 1862–1880").

128. Turner to Christie, 5 Nov. 1897 (RGO 7/195).

129. Hills memo, "Instructions for the Pulgaon eclipse camp," n.d. (RAS Papers 55.3).

130. Herbert Compton, *Indian life in town and country* (London: G. P. Putman's, 1908), chap. 26.

131. Campbell, "In the shadow of the moon," 82.

Chapter 3: The Experience of Fieldwork

1. The evening is described in the (London) *Times*, 10 May 1898, 1. On Lindsay and the Royal Observatory, Edinburgh, see George Forbes, *David Gill, man and astronomer* (London: John Murray, 1916), chap. 5. Lindsay was a generous, if somewhat unusual, patron of science. For the British Museum, for example, he took a crew of naturalists with him on three extended voyages aboard his yacht, the *Valhalla*: see M. J. Nicoll, *Three voyages of a naturalist* (London: Witherby, 1908).

2. "Eclipse slides," list of lantern slides, n.d.; William Christie to Edmond Hills, 27 Apr. 1898 (RGO 7/195).

3. William H. M. Christie, "The photography of the recent total solar eclipse," typed ms., 10 May 1898 (RGO 7/197).

4. There were two dozen eclipses observed from the 1850s to 1914: S. A. Mitchell, *Eclipses of the sun* (New York: Columbia University Press, 1923), 55.

5. On the literature on scientific practice, see Timothy Lenoir, "Practice, reason, context: The dialogue between theory and experiment," *Science in context* 2 (1988), 3–22; Jan Golinski, "The theory of practice and the practice of theory: Sociological approaches in the history of science," *Isis* 81 (1990), 492–505. On thick description, see Clifford Geertz, "Thick description: Toward an interpretive theory of culture," in Geertz, *The interpretation of cultures* (New York: Basic Books, 1973), 3–30.

6. Paul Fussell, *The Great War and modern memory* (Oxford: Oxford University Press, 1975); John Keegan, *The face of battle: A study of Agincourt, Waterloo, and the Somme* (New York: Penguin, 1976); Inga Clendinnen, "The cost of courage in Aztec society," *Past and present* 107 (1985), 44–89; Clendinnen, *Aztecs: An interpretation* (Cambridge: Cambridge University Press, 1989).

7. For a more general discussion on the cultural analysis of travel, see Judith Adler, "Travel as performed art," *American journal of sociology* 94 (1989), 1366–91.

8. On the social and cultural origins of travel conventions, see Malcolm Andrews, *The search for the picturesque: Landscape aesthetics and tourism in Britain* (Palo Alto: Stanford University Press, 1989); Joseph Burke, "The grand tour and the rule of taste," in R. F. Brissenden, *Studies in the 18th century* (Toronto: University of Toronto Press, 1986), 231–50; John Theilmann, "Medieval pilgrims and the origins of tourism," *Journal of popular culture* 20 (1987), 93–102.

9. My analysis is inspired by Lynn Hunt, ed., *The new cultural history* (Berkeley: University of California, 1989); and Alvin Kernan, *Samuel Johnson and the impact of print* (Princeton: Princeton University Press, 1989).

10. (London) *Daily news* clipping, 16 Dec. 1871 (JNL, File "Eclipses 1871, Packet 2").

11. On newspapers and periodicals, see Lucy Brown, *Victorian news and newspapers* (Oxford: Clarendon Press, 1986); Walter Houghton, "Victorian periodical literature and the articulate classes," *Victorian studies* 22 (1979), 389–412; John Sutherland, "*Cornhill*'s Sales and payments: The first decade," *Victorian periodical review* 19 (1986), 106–8. On fiction, see Guinevere L. Griest, *Mudie's circulating library and the Victorian novel* (Bloomington: Indiana University Press, 1970); John Sutherland, *Victorian novelists and publishers* (London: Athlone, 1976); R. C. Terry, *Victorian popular fiction, 1860–1880* (Atlantic Heights, N.J.: Humanities Press, 1983).

12. On scientific journals, see Susan Sheets-Pyenson, "Popular scientific periodicals in Paris and London: The emergence of a low scientific culture, 1820–1875," *Annals of science* 42 (1985), 549–72; Sheets-Pyenson, "A measure of success: The publication of natural history journals in early Victorian Britain," *Publishing history* 9 (1981), 21–36.

13. Charles Pritchard, "A journey in the service of science: Being a description of the phenomena of a total solar eclipse, chiefly as observed at Gujuli, Spain, on July, 18, 1860: Part I," *Good words* (Sept. 1867), 608.

14. Steve Shapin, "Pump and circumstance: Robert Boyle's literary technology," *SSS* 14 (1984), 481–520.

15. A. J. Meadows, *Science and controversy: A biography of Sir Norman Lockyer* (Cambridge: MIT Press, 1972), chaps. 1 and 2; J. J. Thompson, *Recollections and reflections* (London: G. Bell and Sons, 1936), 23–24. British eclipse observers were not the only ones with strong literary interests. In America, Amherst astronomer and eclipse expedition organizer David Todd would have been the doyen of American astronomy textbook-writing had Charles Young—a veteran of eclipse expeditions from 1869 to 1900—not enjoyed immense success with *Elements of astronomy, The sun*, and three other books. His wife, Mabel Loomis Todd, was author of the popular *Total eclipses of the sun.*

16. George Forbes, "The Royal Observatory, Greenwich," *Good words* 13 (1872), 792–96; 855–58; quotations on 793, 795, 855, and 856.

17. Mari Williams, "Astronomical observatories as practical space: The case of Pulkowa," in Frank A. J. L. James, ed., *The development of the laboratory: Essays on the place of experiment in industrial civilization* (New York: American Institute of Physics, 1989), 118–36, esp. 123.

18. "A night at Greenwich Observatory," *Cornhill magazine* 7 (1863), 381–89, esp. 381.

19. Edwin Dunkin, "The Royal Observatory, Greenwich: A day at the observatory," *Leisure hour* 11 (1862), 22–26, quotations on 22, 26.

20. Edwin Dunkin, "The Royal Observatory, Greenwich: A night at the observatory," *Leisure hour* 11 (1862), 55–60, quotations on 55, 58.

21. William Pole, "The eclipse expedition to Spain," *Macmillans* (Sept. 1860), 406–16, quotation on 408.

22. Perry to Lockyer, 11 Jan. 1871 (RAS Papers 51.2); "Lady Astronomer," *Caught in the tropics: A sequel to In pursuit of a shadow* (London: Griffith Farran Okeden and Welsh, 1891), 9.

23. John Tyndall, *Hours of exercise in the Alps* (New York: Appleton and Co., 1897), 431. For other bad voyages, see Mabel Loomis Todd's diary of her expedition to Japan, esp. July 23–28, 1887 (MLT, HM 149, Reel 2).

24. "The English government eclipse expedition," *Nature* (28 Dec. 1871), 164.

25. L. Cumming, "The eclipse expedition," *Nature* (5 Jan. 1871), 184.

26. On the importance of social spaces in scientific productions, see Steven Shapin, "The house of experiment in seventeenth-century England," *Isis* 79 (1988), 369–72.

27. Of course, travel and science have long been interrelated; not only do exploration and the making of new knowledge go hand in hand, but the conventions of early modern travel were strongly influenced by the early scientific academies. See esp. Judith Adler, "Origins of sightseeing," *Annals of tourism research* 16 (1989), 7–29; also Hugh West, "Göttingen and Weimar: The organization of knowledge and social theory in 18th-century Germany," *Central European history* 11 (1978), 150–61.

28. John W. Forney, *A centennial commissioner in Europe, 1874–76* (Philadelphia: J. B. Lippincott, 1876); Joel Cook, *A holiday tour in Europe* (Philadelphia: J. B. Lippincott, 1881); Rev. William Hutton, *12,000 kilometers over land and sea* (Philadelphia: Grant, Faires, and Rodgers, 1878).

29. Daniel Headrick, *The tentacles of progress: Technology transfer in the age of imperialism, 1850–1940* (Oxford: Oxford University Press, 1988), chap. 2.

30. Robert Wiebe, *The search for order* (New York: Hill and Wang, 1967); Alfred Chandler, *The visible hand* (Cambridge: Harvard University Press, 1979); Wolfgang Schivelbusch, *The railway journey* (Berkeley: University of California Press, 1981). On literature and the new technologies, see Christopher Harvie, "'The sons of Martha': Technology, transportation, and Rudyard Kipling," *Victorian studies* 20 (1977), 269–82.

31. Mabel Loomis Todd to E. J. Loomis, June 12, 1887, and June 18, 1887 (DPT, Series II, Box 30, Folder 194). See also Paul Fussell, "Bourgeois travel: Techniques and artifacts," in *Bon Voyage: Designs for travel* (New York: Cooper-Hewitt Museum and Smithsonian Institution, 1986), 55–94.

32. Thomas Cook, quoted in Edmund Swinglehurst, *Cook's tours: The story of popular travel* (Poole: Blandford Press, 1982), 126.

33. F. W. Fairholt, "The Nile as a sanitarium," *The intellectual observer: A review of natural history, microscopic research, and recreative science* 11 (1867), 4.

34. *Grand hotel: The golden age of palace hotels: An architectural and social history* (London: Dent, 1984).

35. Foster R. Dulles, *American abroad: Two centuries of European travel* (Ann Arbor: University of Michigan, 1964); John A. Jakle, *The tourist: Travel in twentieth-century America* (University of Nebraska Press, 1985); C. James Haug, *Leisure and urbanism in nineteenth-century Nice* (Lawrence: Regents Press of Kansas, 1982); Paul Bernard, *Rush to the Alps: The evolution of vaca-*

tioning in Switzerland (New York: Columbia University Press, 1978); Morroe Berger, "Cairo to the American traveler of the nineteenth century," *Colloque international sur L'histoire du Cairo* (Cairo: Ministry of Culture, 1969), 51–65.

36. Edmund Swinglehurst, *Cook's tours: The story of popular travel* (Poole: Blandford, 1982); Piers Brendon, *Thomas Cook: 150 years of popular tourism* (London: Secker and Warburg, 1991).

37. The best study of Victorian travel is John Pemble, *The Mediterranean passion: Victorians and Edwardians in the south* (New York: Oxford University Press, 1987). See also Jonathan Culler, "Semiotics of tourism," *American journal of semiotics* 1 (1981), 127–40.

38. W. S. Caine, *Picturesque India: A handbook for European travelers* (1898; repr. Delhi: Neeraj, 1982); Sidney Low, *A vision of India* (1905; repr. Delhi: Seema, 1975); Edward Said, *Orientalism* (New York: Vintage, 1979), 165–69.

39. Pemble, *The Mediterranean passion*, chaps. 2–6.

40. Ibid., 68–72.

41. Grant Allen, *The European tour: A handbook for Americans and colonists* (London: Grant Richards, 1899); *Baedeker's handbook(s) for travelers: A bibliography of English editions published prior to World War II* (Westport, Conn.: Greenwood Press, 1975). See also E. S. De Beer, "The development of the guide-book until the early nineteenth century," *Journal of the British Archaeological Association* 15 (1952), 35–46. Allen is better known to historians of science as a biologist, Fabian, and advocate of evolutionary arguments for socialism.

42. Eve-Marie Kroller, *Canadian travelers in Europe, 1851–1900* (Vancouver: University of British Columbia Press, 1987), 44.

43. W. Pembroke Feteridge, *Hand-book for travelers in Europe and the East* (New York: Feteridge and Co., 1886), xiv.

44. Kroller, *Canadian travelers in Europe*, 44.

45. W. W. Nevin, *Vignettes of travel* (Philadelphia: J. B. Lippincott, 1881), 438.

46. Pemble, *The Mediterranean passion*, 68–70.

47. "Lady Astronomer" (Miss Brown), *Caught in the tropics: A sequel to In pursuit of a shadow* (London: Griffith Farran Okeden and Welsh, 1891), 46.

48. Pemble, *The Mediterranean passion*, 82–83.

49. John Tyndall, *Hours of exercise in the Alps* (New York: D. Appleton and Co., 1897), 440–42. For an amateur perspective, see Rebecca Joslin, *Chasing eclipses: The total solar eclipses of 1905, 1914, 1925* (Boston: Walton, 1929), 4–9.

50. W. H. M. Christie and H. H. Turner, "Report of the expedition to Sahdol," *RASMN* 58 (1898), 3–4.

51. Rev. Thomas Hincks, "The sea-side," *Student and intellectual observer of science, literature and art* 4 (1870), 167; Rev. R. W. Fraser, *The seaside naturalist* (London: Virtue and Co., 1868); "Field-naturalists' clubs, and how

to form them," *Leisure hour* (6 Apr. 1872), 213–14. John Tyndall and Norman Lockyer were avid mountaineers: See A. S. Eve and C. H. Creasey, *Life and work of John Tyndall* (London: Macmillan, 1945); Meadows, *Science and controversy*. See also David Robbins, "Sport, hegemony and the middle class: The Victorian mountaineers," *Theory, culture and society* 4 (1987), 579–601. On the evolution of leisure in Victorian society, see Peter Bailey, "'A mingled mass of perfectly legitimate pleasures': The Victorian middle class and the problem of leisure," *Victorian studies* 21 (1977), 7–28; Bruce Haley, *The healthy body and Victorian culture* (Cambridge: Harvard University Press, 1977), esp. chaps. 6, 8, and 11; Grace Seibering, *Amateurs, photography, and the mid-Victorian imagination* (Chicago: University of Chicago Press, 1986).

52. Information on the Turner family comes from Leeds censuses of 1861 and 1871, and obituaries of H. H. Turner in the *Yorkshire Post* and *Leeds Mercury*, 21 Aug. 1930. My thanks to Colin Price, Leeds Central Library, for this material.

53. E. W. Maunder, ed., *The Indian eclipse, 1898* (London: BAA, 1899); Maunder, ed., *The total solar eclipse of 1900* (London: Knowledge Office, 1901); F. W. Levander, ed., *The total solar eclipse of 1905* (London: Eyre and Spottiswoode, 1906); E. H. Hills to Ralph Copeland, 22 Feb. 1900 (ROE, 36.264 "Ralph Copeland Correspondence—Letters in 1900"); Hills to Junior Institute of Engineers, 24 Jan. 1905 (RAS Papers 55.2).

54. Roy Porter, "Gentlemen and geology: The emergence of a scientific career, 1660–1920," *Historical journal* 21 (1978), 820. See also Martin Rudwick, *The great Devonian controversy* (Chicago: University of Chicago Press, 1985), 37–41.

55. "The English government eclipse expedition," *Nature* (28 Dec. 1871), 165.

56. J. Norman Lockyer, "The coming eclipse," *Nature* (15 Aug. 1878), 401; George Barker, "On the total solar eclipse of July 29, 1878," *Proc. American Philosophical Society* 18 (1879), 103–4; L. Darwin, A. Schuster, and E. W. Maunder, "Origins of the expedition and general preparation," *Philosophical transactions*, series A, 180 (1889), 291–93; Elizabeth Campbell, "Flint Island eclipse diary," (MLS, Flint Island eclipse Box).

57. J. E. Abbott to Young, 17 Mar. 1898 (CAY, Box 6, File 3).

58. Tennant to Airy, 11, 12, 16, 22 Feb. 1867; Tennant, "Proposal for expeditions to observe the total eclipse of the sun, August 18, 1868," n.d. (RGO 6/122).

59. This description is based on Herbert Compton, *Indian life in town and country* (London: G. P. Putnam's Sons, 1908), chap. 16; and James Morris, *Pax Britannia: The climax of an empire* (New York: Harcourt, Brace, Jovanovich, 1968), 181–88, 266–69.

60. Sir John Strachey, quoted in Sidney Low, *A vision of India* (1906; repr. Delhi: Seema, 1975), 293.

61. This argument follows Thomas Metcalf, "Architecture and the representation of empire: India, 1860–1910," *Representations* 6 (spring 1984), 37–65;

Metcalf, *An imperial vision: Indian architecture and Britain's Raj* (Berkeley: University of California, 1989).

62. Low, *A vision of India*, 295; H. Hervey, *The European in India* (London: Stanley Paul, 1913), 13.

63. See Ronald Inden, *Imagining India* (London: Basil Blackwell, 1990), chap. 4. My thanks to Cheenu Srinivasan for bringing this to my attention.

64. E. Walter Maunder, "The expedition at Talni," in Maunder, ed., *The Indian eclipse 1898*, 11.

65. Information about Guntoor and Buxar is from the *Imperial gazetteer of India: New edition* (Oxford: Clarendon Press, 1909), vol. ix, pp. 247–48, 390–91.

66. "Letter from Col. Tennant to Dr. Huggins," *RASMN* 32 (1872), 70–72; "The total eclipse as seen at Ootacamund," *Nature* (15 Feb. 1872), 300–301. On Ootacamund, see Sir Frederick Price, *Ootacamund: A history* (Madras: Government Printing Office, 1908).

67. Tennant to Lockyer, 29 Nov. 1871 (JNL, eclipse Box 1, File "Eclipse 1871, Packet 1"). For advice to British astronomers traveling to India, see Hester Russell, "Hints for board ship and tent life in India during the eclipse expedition of 1898," *BAA journal* 8 (1897), 38–39.

68. James Anthony Froude, *The English in the West Indies: Or, the bow of Ulysses* (New York: Charles Scribner's Sons, 1900), 53–54.

69. On hotel quality, see *Letters from Jamaica, 'the land of streams and woods'* (Edinburgh: Edmonston and Douglas, 1873), 17–18, 52–53; on Baedekers, see R. Pullen-Burry, *Jamaica as it is, 1903* (London: T. Fisher Unwin, 1903), 4.

70. Quotation is from Pullen-Burry, *Jamaica as it is, 1903*, 44; see also *Letters from Jamaica, "the land of streams and woods"* (Edinburgh: Edmonston and Douglas, 1873), 54.

71. Darwin, Schuster, and Maunder, "Origins of the expedition and general preparation," 293; H. H. Turner, "Report of the observations of the total solar eclipse of August 29, 1886, made at Grenville, in the Island of Grenada," *Philosophical transactions*, series A, 180 (1880), 385; S. J. Perry, "Report of the observations of the total solar eclipse of August 29, 1886, made at the Island of Carriacou," *Philosophical transactions*, series A, 180 (1889), 352.

72. "Lady Astronomer" (Miss Brown), *Caught in the tropics*, chaps. 6–8.

73. "The Australian eclipse expedition," *Nature* (29 Feb. 1872), 351.

74. H. H. Turner, draft of article or speech, n.d. (JNL, eclipse Box 2, File "Eclipse 1886").

75. Lawrance to Lockyer, 14 June 1883 (JNL, eclipse Box 1, File "Eclipse 1883").

76. *Reports of the observations of the total eclipse of the sun, December 21–22, 1889* (Sacramento: State Printing Office, 1891), 1–2, 23; "The eclipse expedition," *The times* (26 Dec. 1870), 8; J. N. Lockyer, "The Mediterranean eclipse, Part I," *Nature* (19 Jan. 1871), 222; Lockyer, "The Mediterranean eclipse, 1870, Part II," *Nature* (16 Feb. 1871), 321; "The Australian eclipse expedition," *Nature* (29 Feb. 1872), 351.

77. *Reports of the observations of the total eclipse of the sun, August 7, 1869* (Washington: Government Printing Office, 1870), plates 1 and 2; "The eclipse observations at Bekul," *Nature* (Feb. 1872), 265–67; "The Approaching eclipse," *Nature* (12 Apr. 1883), 556; J. Norman Lockyer, *Recent and coming eclipses: Being notes on the total solar eclipses of 1893, 1896, and 1898* (London: Macmillan and Co., 1897), frontispiece and illustrations 18, 27, and 56; Mabel L. Todd, *Total eclipses of the sun* (Boston: Roberts Brothers, 1894), 182, 211.

78. Charles Young, "Observations upon the solar eclipse of July 29, 1878, by the Princeton eclipse expedition," 279; *Report of the observations of the total eclipse of the sun, published by the Lick Observatory, January 1, 1889* (Sacramento: State Printing Office, 1889), 74, 125.

79. Turner, "Report of the eclipse committee of the Royal Astronomical Society, 1890 March 14," *RASMN* 50 (1890), 272–73; Lockyer, "The total eclipse of the sun, April 16th, 1893," *Philosophical transactions*, series A, 187 (1896), quotation on 567.

80. H. H. Turner, draft of article or speech, n.d. (JNL, eclipse Box 2, File "Eclipse 1886").

81. Turner to E. B. Knobel, 22 Mar. 1889 (RAS Papers 53.2).

82. In addition to sources below, see also Chicago Astronomical Society, *The solar eclipse of July 29, 1878* (Chicago: Evening Journal Book and Job Printing House, 1878), 13; C. J. Joly et al., "The total solar eclipse of 1900," *Transactions of the Royal Irish Academy*, section A, 32 (1903), 271–98, on 273.

83. Warren De la Rue, "On the total solar eclipse of July 18th, 1860, observed at Rivabellosa, near Miranda de Ebro, in Spain" (Bakerian Lecture 1862), *Philosophical transactions* 158 (1862), quotation on 353–54.

84. William Pole, "The eclipse expedition to Spain," *Macmillans* (Sept. 1860), 406–16.

85. Some observatory planners selected sites away from cities for the same reasons, to stem to flow of visitors and foster a local culture; see Charles Young, "An astronomer's summer trip," *Scribner's magazine* (July 1888), esp. 90–91; Mari E. W. Williams, "Astronomical observatories as practical space: The case of Pulkowa," in Frank A. J. L. James, ed., *The development of the laboratory: Essays on the place of experiment in industrial civilization* (New York: American Institute of Physics, 1989), 118–36; Donald Osterbrock, John R. Gustafson, W. J. Shiloh Unruh, *Eye on the sky: Lick Observatory's first century* (Berkeley: University of California Press, 1988), passim.

86. Lockyer to Darwin, 17 Aug. 1886; Darwin to Lockyer, 17 Aug. 1886 (JNL, eclipse Box 2, File "Eclipse 1886"); *Rocky Mountain news* (28 July 1878), 4; David Todd, 1887 eclipse diary, pp. 43, 48, 55 (DPT, Series II, Box 30, Folder 198); Elizabeth Campbell, "In the shadow of the moon: India, January 22, 1898," unpublished ms. (MLS), 101, 119–28; Elizabeth Campbell, 1905 eclipse diary (MLS, Spanish eclipse File 6), 65, 69–71, 85–89. As an 1863 article on Greenwich explained, "If . . . we knock at the door and ask permis-

sion to view the establishment, we shall be politely informed that 'no visitors are admitted.' This seclusion is absolutely necessary, for the staff is a hard-worked one, and is not to be interrupted. The instruments are most delicate, and a touch—the resting of a hand on a screw or lever, or even breathing on a portion which is likely to rust—might cause damage or delay, which could not possibly be afforded. In long calculations, perfect quiet is also necessary": "A night at Greenwich Observatory," *Cornhill magazine* 7 (1863), 381–89, quotation on 381.

87. Tyndall, *Hours of exercise in the Alps* (New York: Appleton and Co., 1897), 444–46; *Report of the observations of the total eclipse of the sun, published by the Lick Observatory, January 1, 1889* (Sacramento: State Printing Office, 1889), 33–35, 57–59.

88. Edmond Grove-Hills, "Total solar eclipse of 1901 May 17: Assistance required from the survey department of India," n.d. (RAS Papers 55.2). A similar change took place with naval officers: in the 1860s, they were co-observers with astronomers, but by the 1890s they and their crews were easy sources of labor. J. Norman Lockyer to Edmond Grove-Hills, ? Feb. 1900 (RAS Papers, 55.2).

89. Simon Newcomb diary, 23–28 July 1878, Brown notebook #833 (SNP, Box 1, "Diaries and Notebooks"); S. J. Perry, "Report of the observations of the total solar eclipse of August 29, 1886, made at the Island of Carriacou," *Philosophical transactions,* series A, 180 (1889), 356–57; David Todd, "Adjustment of the photographic telescope"; R. Hitchcock, "Report of operations in the photographic house," entries for 23 July–18 Aug. 1887 (DPT, Series II, Box 31, Folder 201); Campbell, "In the shadow of the moon," 124–25.

90. Henry Davis, "Addenda to 'notes on eclipse photography,'" handwritten ms., n.d. (JNL, eclipse Box 1, File "Eclipse 1871, Packet 2").

91. Lockyer, "The English eclipse expedition," *Nature* (18 May 1882), 63–65, quotation on 63.

92. Mary Louise Pratt, *Imperial eyes: Travel writing and transculturation* (London: Routledge, 1992).

93. Eben J. Loomis, *An eclipse party in Africa: Chasing summer across the equator in the USS Pensacola* (Boston: Roberts Brothers, 1896), 57, 58.

94. Elizabeth Campbell, "In pursuit of a shadow: Spain 1905" (MLS, Box "Eclipses, Spain 1905," File 6), 87a.

95. Henry Cousens, "The camp at Jeur," in E. Walter Maunder, ed., *The Indian eclipse, 1898* (London: Hazell, Watson and Viney, 1899), chap. 6, quotation on 50.

96. Ibid., 93–94.

97. Elizabeth Campbell, "In pursuit of a shadow: Spain 1905," 87a.

98. Campbell, "In the shadow of the moon: India, January 22, 1898," 123.

99. James Tennant to Airy, 16 July 1868 (RGO 6/122).

100. Arthur Schuster, *Biographical fragments* (London: Macmillan and Co., 1932), 89.

101. "Lady Astronomer" (Miss Brown), *Caught in the tropics,* 72.

102. "Lady Astronomer" (Miss Brown), *In pursuit of a shadow*, 103.

103. For example, one amateur describes the arrival of science students "of rather military appearance," being dismissed because "our astronomers evidently thought they had come too late to be of much use, not having gone through [her companion] L.'s severe training." "Lady Astronomer" (Miss Brown), *In pursuit of a shadow*, 108.

104. E. W. Maunder, "The total eclipse of August 9, 1896," *Knowledge*, Sept. 1896, 204. On the negotiation of order in the workplace, see Michael Burawoy, *Manufacturing consent: Changes in the labor process under monopoly capital* (Chicago: University of Chicago Press, 1979); Merritt Roe Smith, *Harper's Ferry Armory and the new technology* (Ithaca: Cornell University Press, 1977).

105. Simon Schaffer, "Astronomers mark time: Discipline and the personal equation," *Science in context* 2 (1988), 115–45.

106. A private observatory was a symbol of the investment of time, money, and energy that an amateur was willing and able to make; it separated the serious amateur from the casual observer; Frederick Bird, "A cheap observatory," *The intellectual observer* 6 (1964–65), 242; L. Rudaux, *How to study the stars* (London: T. Unwin, 1909), 127; Rev. N. S. Heineken, "Substitutes for an observatory," *The intellectual observer* 6 (1864–65), 444–46.

107. On the Franklin Expedition, see Owen Beattie and John Geiger, *Frozen in time: The fate of the Franklin expedition* (London: Bloomsbury, 1987), chaps. 2–5; Leslie Neatby, *The search for Franklin* (New York: Walker and Co., 1970). On Burton-Speke, see Edward Rice, *Captain Sir Richard Francis Burton* (New York: Charles Scribner's Sons, 1990), 260–313; see also James Casada, *Sir Richard Francis Burton: A biobibliography* (Burton, Mass.: G. K. Hall, 1990). Average lifespan is computed in Richard van Orman, *The explorers: Nineteenth century expeditions in Africa and the American west* (Albuquerque: University of New Mexico Press, 1984), 159–65; John Keay, *When men and mountains meet: The explorers of the western Himalayas, 1820–1875* (London: John Murray, 1977), 107–21, chap. 12. John Krakauer, *Into thin air: A personal account of the Mt. Everest disaster* (New York: Villard, 1997), is a brilliant account of making life-or-death decisions in the field.

108. Henry Walter Bates, *The naturalist on the River Amazons* (1863; repr. New York: Penguin Books, 1989); Charles Darwin, *The voyage of the Beagle* (New York: Mentor, 1988); William Goetzmann, *New lands, new men: America and the second great age of discovery* (New York: Penguin Books, 1986), chap. 7.

109. Arthur Schuster spent several months after the eclipse of 1875 touring the Himalayas; numerous American astronomers combined expeditions with European tours. See Schuster, *Biographical fragments* (London: Macmillan and Co., 1932), chaps. 10–11; Simon Newcomb, *Reminiscences of an astronomer* (Cambridge: Houghton Mifflin, 1903), chaps. 20–21.

110. Edward Pickering, "The eclipse of 1870," *Old and new* 3 (1871), 634. S. A. Mitchell explained that "first contact is difficult to observe with accuracy

since nothing is to be seen at the edge of the sun until the moon is actually projected onto the face of the sun—and first contact has already actually taken place." Mitchell, *Eclipses of the sun,* 66.

111. R. F. Chisholm, "Extracts from general observations on the total eclipse of December 12th, 1871," *Astronomical register* 10 (1872), 237–39, esp. 237.

112. Hermann Vogel, "The photographic expedition to Aden, Arabia," *Philadelphia photographer* 5 (1868), 365; see also Frederic Mylius, "Pictures of the eclipse," *Camera craft* (July 1900), 112–15.

113. Henry Holiday, *Reminiscences of my life* (London: William Heinemann, 1914), 210.

114. Warren De la Rue, "On the total solar eclipse of July 18th, 1860," 353.

115. William Pole, "The eclipse expedition to Spain," 413.

116. C. F. Himes, "The total eclipse of the sun of 1869," *American journal of science* (1869), 145.

117. J. Norman Lockyer, "The total solar eclipse of August 9, 1896—Report on the expedition to Kiö Island," *Philosophical transactions* 190 (1897), 1–21, quotation on 20.

118. Ben Hur Wilson, "The total eclipse of 1869," *The palimpsest* 51 (1970), 92.

119. Chicago Astronomical Society, *The solar eclipse of July 29, 1878* (Chicago: Evening Journal Book and Job Printing House, 1878), 40.

120. J. J. Aubertin, *By order of the sun to Chile to see his total eclipse April 16, 1893* (London: Kegan Paul, Trench Trübner and Co., 1894), 56.

121. "Lady Astronomer" (Miss Brown), *Caught in the tropics,* 17–18, 33, 35; Aubertin, *By order of the sun to Chile to see his total eclipse April 16, 1893,* esp. 7–8, 17, 41–43.

122. "The Australian eclipse expedition," *Nature* (29 Feb. 1872), 351–54; H. O. Russell, "Eclipse of December last—Australian expedition," *RASMN* (Mar. 1872), 220–21.

123. A. A. Common, "Report on the Vadsö expedition," *RASMN* 47 (1897), 115.

124. "Lady Astronomer" (Miss Brown), *In pursuit of a shadow,* 111.

125. "The eclipse that failed," *Daily telegraph,* 19 Aug. 1896 (clipping in JNL, Box 1, File "1896 eclipse Vadsö").

126. This was first brought to my attention by H. Schellen, *Spectrum analysis in its application to terrestrial substances, and the physical constitution of the heavenly bodies,* trans. Jane and Caroline Lassell, ed. William Huggins (London: Longmans, Green and Co., 1872), 298; see also Loomis, *An eclipse party in Africa,* 70.

127. Rev. Charles Pritchard, "A journey in the service of science," *Good words* (Oct. 1867), 698. Italics in the original.

128. Pole, "The eclipse expedition to Spain," 413.

129. Edward Pickering, "The eclipse of 1870," *Old and new* 3 (1871), 634. See also Joslin, *Chasing eclipses,* 13. This was often a more accurate measure of

second contact than observation: see *Reports of observations of the total eclipse of the sun, August 7, 1869*, 15.

130. Aubertin, *By order of the sun to Chile to see his total eclipse*, 141.

131. Chisholm, "Extracts from general observations on the total eclipse of December 12th, 1871," *Astronomical register* 10 (1872), 237.

132. "The eclipse," *New York Times* (12 Aug. 1869), 4.

133. John P. Maclear, "Captain Maclear's observations," *Nature* (18 Jan. 1872), 220–21; Thomas Milner, *The heavens and the earth*, 56; "The sun's corona," *Cornhill magazine* (Oct. 1870), 482; E. Colbert, "Light from the eclipse," *Lakeside monthly* 5 (Mar. 1871), 228.

134. Lockyer, "The English eclipse expedition," repr. in Lockyer, *Contributions to solar physics* (London: Macmillan and Co., 1874), 334.

135. Francis Galton, "A visit to north Spain at the time of the eclipse," in Galton, ed., *Vacation tourists and notes of travel in 1860* (London: Macmillan and Co., 1861), 437.

136. *Burlington hawkeye* (10 Aug. 1869), 2.

137. Lockyer, "The total eclipse," *Nature* (18 Jan. 1872), 218.

138. Pritchard, "A journey in the service of science" (part 1), *Good words* (Sept. 1867), 608.

139. George Airy, "Suggestions to astronomers for the observation of the total eclipse of the sun on 1851 July 28," draft ms., 13 Dec. 1850 (RGO 6/115).

140. De la Rue, "On the total solar eclipse of July 18th, 1860," 356. A similar reaction is described by Major James F. Tennant, "Report of the total eclipse of the sun, August 17–18, 1868," *RAS memoirs* 37 (1869), 19.

141. Charles Piazzi Smyth, "On the total solar eclipse of 1851," *Transactions of the Philosophical Society of Edinburgh* 20 (1852), 504.

142. R. C. Carrington, *Information and suggestions addressed to persons who may be able to place themselves within the shadow of the total eclipse of the sun* (London: Eyre and Spottiswoode, 1858), 18.

143. Carrington to Airy, 27 Apr. 1858 (RGO 6/121).

144. Andrews, *The search for the picturesque*; Ian Ousby, *The Englishman's England: Taste, travel and the rise of tourism* (Cambridge: Cambridge University Press, 1990).

145. Descriptions of the onset of totality are brief and contain none of the sublime imagery of the 1860s in Leonard Darwin et al., "On the total solar eclipse of August 29, 1886," *Philosophical transactions*, series A, 180 (1889), 296; and H. H. Turner, "Report of eclipse committee of the RAS," *RASMN* 50 (1890), 273, 279. Accounts that only mention the times of contact include S. W. Burnham and J. M. Schaeberle, *Reports on the observations of the total eclipse of the sun* (Sacramento: Lick Observatory, 1891), J. Norman Lockyer, "The total eclipse of the sun, April 16, 1893," *Philosophical transactions*, series A, 187 (1896), 553–606.

146. See accounts in *The Indian eclipse* (BAA, 1899) and *RASMN* 58 (1898), 6, 25.

147. Ralph Copeland, "Total solar eclipse of 21 January 1898," undated ms., 17 (ROE 36.300).

148. Campbell, "In the shadow of the moon: India, January 22, 1898," 130.

149. Ibid., 2.

150. Arthur Schuster, *Biographical fragments* (London: Macmillan and Co., 1932), 91.

151. Smyth, "On the total solar eclipse of 1851," 507.

152. Young, "An astronomer's summer trip," 82–100, esp. 95.

153. George Frederick Chambers, *The story of eclipses, simply told for general readers* (London: Hodder and Stoughton, 1902), quotation on 228–29. On African indigenous responses, see also Sir Robert Stawell Ball, *In the high heavens* (London: Isbister and Co., 1901), 80.

154. Quoted in Todd, *Total eclipses of the sun*, 141.

155. *Denver tribune* (30 July 1878), 4; *Rocky Mountain news* (30 July 1878), 4.

156. Turner, draft of talk, n.d. (JNL, eclipse Box 2, File "Eclipse 1886").

157. Charles Pritchard, "A journey in the service of science: Being a description of the phenomena of a total solar eclipse, chiefly as observed at Gujuli, Spain, on July 18, 1860" (part 2), *Good words* (Oct. 1860), 694–701, quotation on 695.

158. Alfred Brothers, "Eclipse expedition 1870, Sicilian section, photographic department," n.d. (RAS, Ranyard Papers, MSS 5.2).

159. Young, "An astronomer's summer trip," 82–100, esp. 95.

160. Edward J. Stone, "Observations of the total solar eclipse of April 16, 1874, at Klipfontein, Namaqualand, South Africa," *RASMN* 34 (June 1874), 401.

161. Lockyer, "The eclipse expedition," *Nature* (8 June 1882), 129–30, quotation on 129.

162. John P. Maclear, "Captain Maclear's observations," *Nature* (18 Jan. 1872), 219–21, quotation on 221.

163. J. Norman Lockyer, "The solar eclipse," *Nature* (1 Feb. 1872), 260.

164. J. Boesinger, "The total eclipse as seen at Ootacamund," *Nature* (15 Feb. 1872), 300–301, quotation on 301.

165. "Total solar eclipse, January 22, 1898," *Knowledge* (Feb. 1898), 38–39, esp. 39.

166. Lockyer, *Contributions to solar physics* (London: Macmillan and Co., 1874), 343–44.

167. Chambers, *The story of eclipses*, 224; Todd, *Total eclipses of the sun*, 81.

168. Lockyer, *Recent and coming eclipses*, 18.

169. David Edwards, "Mad mullahs and Englishmen: Discourse in the colonial encounter," *CSSH* 31 (1989), 649–70.

170. Brian Bond, ed., *Victorian military campaigns* (London: Hutchinson, 1967). For more on technology and imperialism, see Daniel Headrick, *The tools of empire: Technology and European imperialism in the nineteenth cen-*

tury (Oxford: Oxford University Press, 1981); on Cook's voyages, Bernard Smith, *European vision and the South Pacific* (New Haven: Yale University Press, 1985). On science and imperialism, see Lewis Pyenson, *Cultural imperialism and exact science: German expansion overseas, 1900–1930* (New York: Peter Lang, 1983); Robert Stafford, "Geological surveys, mineral discoveries, and British expansion, 1835–71," *Journal of imperial and commonwealth history* 12 (1984), 5–32.

171. Norman Pogson to George Airy, 29 Mar. 1871 (RGO 6/134).

172. De la Rue, "On the total solar eclipse of July 18th, 1860, observed at Rivabellosa, near Miranda de Ebro, in Spain," quotation on 353–54.

173. This explanation is related in a letter from J. F. Tennant, *Observatory* 19 (1896), 276–77; letter from C. Michie Smith, ibid., 334–35. The letters were a response to (and in fact a refutation of) "From an Oxford notebook," *Observatory* 19 (1896), 252, an account of "Hindoo" eclipse superstitions.

174. Violet Jacob, *Diaries and letters from India, 1895–1900,* ed. Carol Anderson (Edinburgh: Canongate, 1990), 110.

Chapter 4: Drawing and Photographing the Corona

1. This section draws upon my "Visual representation and post-constructivist history of science," *HSPS* 27 (1997), 139–71. Other reviews include Jan Golinski, "The theory of practice and the practice of theory: Sociological approaches in the history of science," *Isis* 81 (1990), 492–505; Andrew Pickering, "From science as knowledge to science as practice," in Pickering, ed., *Science as practice and culture* (Chicago: University of Chicago Press, 1992), 1–26.

2. See, for example, Andrew Pickering, "Beyond constraint: The temporality of practice and the historicity of knowledge," in Jed Buchwald, ed., *Scientific practice: Theories and stories of doing physics* (Chicago: University of Chicago Press, 1995), 42–55.

3. Philip Gilbert Hamerton, *The graphic arts: A treatise on varieties of drawing, painting, and engraving in comparison with each other and with nature* (Boston: Roberts Brothers, 1882), 1–2.

4. Jay David Bolter, *Writing space: The computer, hypertext, and the history of writing* (Hillsdale, N.J.: Lawrence Erlbaum Associates, 1991); Alvin Kernan, *The death of literature* (New Haven: Yale University Press, 1990).

5. The limits of literary criticism in art history have also been argued by Barbara Stafford in *Body criticism: Imaging the unseen in Enlightenment art and medicine* (Cambridge: MIT Press, 1991); and Stafford, "Presuming Images and Consuming Words: The Visualization of Knowledge from the Enlightenment to Post-Modernism," in John Brewer and Roy Porter, eds., *Consumption and the World of Goods* (London: Routledge, 1993), 462–77.

6. Philip Gilbert Hamerton, *The graphic arts: A treatise on varieties of drawing, painting, and engraving in comparison with each other and with nature* (Boston: Roberts Brothers, 1882), 1–2.

7. My thinking on the topic of visual representation in astronomy has been

influenced by Samuel Y. Edgerton, "Galileo, Florentine 'disegno,' and the 'strange spottednesse' of the moon," *Art journal* 44 (1984), 225–32; William B. Ashworth, *The face of the moon: Galileo to Apollo* (exhibit catalog, Linda Hall Library, Kansas City, Mo., 13 Oct. 1989–28 Feb. 1990); Edgerton and Michael Lynch, "Aesthetics and digital image processing: Representational craft in contemporary astronomy," in Gordon Fyfe and John Law, eds., *Picturing power: Visual depiction and social relations* [Sociological Review Monograph 35] (London: Routledge, 1988), 184–219. On imaging practices in the nineteenth century, I have especially drawn on Ann Shelby Blum, *Picturing nature: American nineteenth-century zoological illustration* (Princeton: Princeton University Press, 1993); Lorraine Daston and Peter Galison, "The image of objectivity," *Representations* 40 (fall 1992), 81–128; Holly Rothermel, "Images of the sun: Warren De la Rue, George Biddell Airy and celestial photography," *BJHS* 26 (1993), 137–69; Simon Schaffer, "Astronomers mark time: Discipline and the personal equation," *Science in context* 2 (1988), 115–45.

8. Steven Shapin, "The invisible technician," *American scientist* 77 (1989), 554–63. The advantages of studying unusual and contentious periods in the history of science have been outlined by students of scientific controversy: see Harry Collins, *Changing order: Replication and induction in scientific practice* (Beverly Hills: Sage, 1985); Trevor Pinch, *Confronting nature: The sociology of solar-neutrino detection* (Dordrecht: D. Reidel, 1986); Martin Rudwick, *The great Devonian controversy: The shaping of scientific knowledge among gentlemanly specialists* (Chicago: University of Chicago Press, 1985); Steven Shapin and Simon Schaffer, *Leviathan and the air-pump: Hobbes, Boyle, and the experimental life* (Princeton: Princeton University Press, 1985).

9. On the place of eclipse observation in solar physics, see Karl Hufbauer, *Exploring the sun: Solar science since Galileo* (Baltimore: Johns Hopkins University Press, 1991).

10. On technology and amateur-professional debates, see John Lankford, "Amateur versus professional: The transatlantic debate over the measurement of Jovian longitude," *BAA journal* 89 (1979), 574–82; and John Lankford, "Amateurs versus professionals: The controversy over telescope size in late Victorian science," *Isis* 72 (1981), 11–28.

11. On the mid-Victorian drawing-photography relationship, see Philip Gilbert Hamerton, "Relation between photography and painting," in Philip Gilbert Hamerton, *Thoughts about art* (London: Macmillan and Co., 1889), chap. 4; Grace Seibering, *Amateurs, photography, and the mid-Victorian imagination* (Chicago: University of Chicago Press, 1986); Doug Nickel, "The camera and other drawing machines," in Mike Weaver, ed., *British photography in the nineteenth century: The fine art tradition* (Cambridge: Cambridge University Press, 1989), 1–10.

12. John Brett to Lockyer, 18 Nov. 1870 (RAS Papers 51.1). On Brett, see Allen Staley, *The Pre-Raphaelite landscape* (Oxford: Oxford University Press, 1973), 124–37; Kenneth Bendiner, "John Brett's 'The glacier of Rosenlaui,'" *Art*

journal 44 (1984), 241–48; on Holiday, see Henry Holiday, *Reminiscences of my life* (London: W. Heinemann, 1914).

13. The 1860 eclipse was drawn by Joseph Bonomi, an illustrator whose work was "scattered through all the principal Egyptologists' publications of his time": "Bonomi, Joseph the younger," *Dictionary of National Biography* 5, 364. See also Samuel Hunter to Arthur Ranyard, 16 Nov. 1870 (RAS Papers 51.2); "Report by Captain A. B. Fyers, R. E. on the Eclipse of 1871," n.d. (RAS, Ranyard Papers, MSS 5.4).

14. Charles Piazzi Smyth, "On astronomical drawing," *Memoirs RAS* 15 (1846), 71–82, quotation on 75.

15. John Hershel, *Outlines of astronomy*, 4th ed. (Philadelphia: Blanchard and Lea, 1860), 512.

16. John Brett, "Instructions to observers," handwritten ms., 18 Nov. 1870 (RAS Papers 51.1); Brett, "Some particulars to be especially noticed by those observers who make drawings of the corona," handwritten ms., n.d. (JNL, File "Eclipse 1871—Packet 2").

17. Chicago Astronomical Society, *The solar eclipse of July 29, 1878* (Chicago: Evening Journal Book and Job Printing House, 1878), 11. Copy in the University of Pennsylvania Rare Book Room.

18. For extended descriptions of drawing practices, see S. Devenish to Airy, 6 Jan. 1862 (RGO 6/115); C. E. Burton to Arthur C. Ranyard, n.d. 1871 (RAS, Ranyard Papers, MSS 5.4); James Carpenter, "Report on proceedings in reference to the solar eclipse expedition, 1870 December," handwritten ms., 12 Jan. 1871 (RGO 6/131), 12–13, 16–17; Chisholm, "Extracts from general observations on the total eclipse of December 12, 1871," 237–39; Holiday, *Reminiscences of my life* (cit. n. 11), 170, 171, 209, 211; on materials, see Macdonald, "Report of the eclipse of the 18th August 1868," 215; Stone, "Observations of the total solar eclipse of April 16, 1874," 399–401; Stone, "Solar eclipse without instrumental means," 451–52; on blindfolding, see George Airy, "Suggestions to astronomer for the observation of the total eclipse of the sun on 1851 July 28," handwritten ms., n.d. (RGO 6/119); J. R. Hind, "Instructions on the total eclipse of the sun, 31 December 1861," handwritten ms., n.d. (RGO 6/115); Arthur Schuster, "Some remarks on the total solar eclipse of July 29, 1878," *RASMN* 39 (1878), 45; Simon Newcomb diary, 29 July 1878, Brown notebook no. 833 (SNP, Box 1, "Diaries and Notebooks").

19. Smyth, "On astronomical drawing," 75, 72.

20. Chisholm, "Extracts from general observations on the total eclipse of December 12, 1871," 238; Holiday, *Reminiscences of my life*, 210, 211. On the emotional challenges of eclipse observation, see Alex Soojung-Kim Pang, "The social event of the season: Solar eclipse expeditions and Victorian culture," *Isis* 84 (June 1993), 252–77.

21. Herschel quoted in "The great eclipse of the sun, part 1," *The engineer* (6 Nov. 1868), 345–46.

22. Airy to Tennant, 5 Oct. 1868 (RGO 6/122).

23. Francis Galton, "A visit to north Spain at the time of the eclipse," in Galton, ed., *Vacation tourists and notes of travel in 1860* (London: Macmillan and Co., 1861), 440; Warren De la Rue, "On the total solar eclipse of July 18th, 1860, observed at Rivabellosa, near Miranda de Ebro, in Spain" (Bakerian Lecture 1862), *Philosophical transactions* 158 (1862), 356; James Tennant, "Report on the total eclipse of the sun, August 17–18, 1868," *Memoirs RAS* 37 (1869), 21; Chisholm, "Extracts from general observations on the total eclipse of December 12, 1871," 239; Captain Tanner, "Remarks on the total eclipse of the 18th of August, 1868," *Proceedings of the Asiatic Society of Bengal* (Sept. 1868), 212; George Darwin to Arthur C. Ranyard, n.d. (RAS, Ranyard Papers, MSS 5.2); James F. Tennant, "Suggestions for visitors to the total eclipse of 12 December 1871," *Proceedings of the Asiatic Society of Bengal* (July 1871), 150–55, quotation on 154.

24. Helmut Gernsheim, *A concise history of photography* (Mineola: Dover, 1986), 16.

25. Mabel Loomis Todd, *Total eclipses of the sun* (Boston: Roberts Bros., 1894), chap. 11; L. Darwin, Arthur Schuster, and E. W. Maunder, "Preparing for the Expedition at Prickly Point," *Philosophical transactions*, series A, 180 (1889), 293.

26. Ranier Fabian and Hans-Christian Adam, *Masters of early travel photography* (New York: Vendome, 1988), 16–19.

27. Smyth, "On Astronomical Photography," 72. Again, Smyth's views follow those of John Herschel, who described the problem of representations of nebulae as originating "partly from the difficulty of correctly drawing, and still more, engraving such objects": Herschel, *Outlines of Astronomy* (London: Longman, Brown, Green and Longmans, 1849), 512.

28. Richard Proctor, "First fruits of the eclipse observations," *Gentleman's magazine* (Sept. 1878), 299. Italics in the original.

29. Making positives from reproduced negatives was standard practice, as it protected the original: see Charles Smyth, "On photographic illustrations for books," *Transactions of the Scottish Society of Arts* 5 (1861), 87–92; Warren De la Rue, "On the total solar eclipse of July 18th, 1860" (Bakerian Lecture 1862), *Philosophical transactions* 158 (1862), 399. On retouching, see Alfred Brothers to Norman Lockyer, 4 Feb. 1871 (RAS Papers 51.1); Ranyard to Airy, 30 Jan. 1872 (RGO 6/133).

30. The literature on engraving and photomechanical reproduction is immense. For historians of Victorian science, essential works include William Ivins, *Prints and visual communication* (Cambridge: Harvard University Press, 1953; repr. MIT Press, 1973); Gavin Bridson and Geoffrey Wakeman, *Printmaking and picture printing: A bibliographical guide to artistic and industrial techniques in Britain 1750–1900* (Williamsburg, Va.: Bookpress, 1984); Bamber Gascoigne, *How to identify prints: A complete guide to manual and mechanical processes from woodcut to ink-jet* (New York: Thames and Hudson, 1986).

31. This discussion is based on Anthony Dyson, *Pictures to print: The nine-*

teenth-century engraving trade (London: Farrand Press, 1984); Celina Fox, "The engravers' battle for professional recognition in early nineteenth-century London," *London journal* 2 (1976), 3–32; Estelle Jussim, *Visual communication and the graphic arts: Photographic technologies in the nineteenth century* (New York: Bowker, 1983), chap. 8; Lucien Goldschmidt, "Tangible fact, poetic interpretation," in Goldschmidt and Weston J. Naef, *The truthful lens: A survey of the photographically illustrated book, 1844–1914* (New York: Grolier Club, 1980), 3–7.

32. As one writer put it, at the height of his craft the best engraver "ceases to be a copyist to become a translator": Charles Blanc, *Grammar of painting and engraving* (Cambridge: Riverside Press, 1874), 245.

33. Estelle Jussim, "The syntax of reality," in *The eternal moment: Essays on the photographic image* (New York: Aperture, 1989), 19–36; Jussim, *Visual communication and the graphic arts*, chap. 9.

34. Holly Rothermel, "Images of the sun," quotation on 153.

35. Airy to De la Rue, 27 Aug. 1860 (RGO 6/123).

36. Airy to De la Rue, 27 Mar. 1861 (RGO 6/123).

37. De la Rue to Airy, 28 Mar. 1861 (RGO 6/123).

38. Tennant to Lockyer, 9 Sept. 1872 (JNL, Lockyer Letters, File T).

39. Warren De la Rue, "On heliotypography," *RASMN* 22 (1862), 278–79; De la Rue, "On a photo-engraving of a lunar photograph," *RASMN* 25 (1865), 171.

40. Helena E. Wright, *Imperishable beauty: Pictures printed in collotype* (Catalog of exhibition, National Museum of American History, Jan. 1988); Alfred Walter Elson, *Lectures on printing* (Cambridge: Harvard University Press, 1913), part iv, 35–37.

41. Elson, *Lectures on printing*, 36–37, quotation on 36; De la Rue to Airy, 17 July 1869 (RGO 6/122).

42. Philip Gilbert Hamerton, *Thoughts about art* (1873; repr. London: Macmillan and Co., 1889), 58.

43. James F. Tennant, "Suggestions for visitors to the total eclipse of 12 December 1871," *Proceedings of the Asiatic Society of Bengal* (July 1871), 150–55, quotation on 154.

44. De la Rue discovered an interesting variation in the brightness of prominences by making "copies obtained by a very long exposure of the original negative which rings out details [normally] masked": De la Rue to Airy, 28 Mar. 1861 (RGO 6/123).

45. Major John F. Tennant, "Report of the total eclipse of the sun, August 17–18, 1868," 35.

46. Alfred Brothers, "Eclipse expedition 1870, Sicilian Section, Photography Department," handwritten ms., n.d., n.p. (RAS MSS 5.2). This attitude gave Brothers license to alter his photographs: he painted in the prominences on his glass positive copies of the 1870 eclipse. See glass photograph "Brothers, Sicily, December 1870" (CAY, Box 3, Folder 31, "Miscellaneous Folders").

47. H. Lawrance and C. Ray Woods to Lockyer, 21 May 1883 (JNL, File

"Eclipse 1883"). Lawrance and Woods wrote from Caroline Island, in the South Pacific.

48. Warren De la Rue, "On the total solar eclipse of July 18th, 1860," *Philosophical transactions*, 409. This also probably created problems for engravers, who were taught to ignore detail and instead use certain methods to suggest softness, transparency, and so forth: Charles Blanc, *Grammar of painting and engraving* (Cambridge: Riverside Press, 1874), chap. 3. Lynn Merrill likewise argues that lithography was popular in natural history illustration because "it suggests soft textures and fluid curves" and was well suited to depicting furry and feathery animals, always more popular subjects than crustaceans and insects: Lynn Merrill, *The romance of Victorian natural history* (Oxford: Oxford University Press, 1989), 166.

49. Charles Piazzi Smyth, *Tenerife, an astronomer's experiment: Or specialties of a residence above the clouds* (London: Lovell Reeve, 1858), 426. On the negotiation of authority between scientists and illustrators, see Alex Soojung-Kim Pang, "'Stars should henceforth register themselves': Astrophotography at the Lick Observatory," *BJHS* 30 (1997), 177–202; and Brian Dolan, "Pedagogy through print: James Sowerby, John Mawe and the problem of color in early nineteenth-century natural history illustration," *BJHS* 31 (1998), 275–304.

50. Arthur C. Ranyard, ed., "Observations made during total solar eclipses," *RAS Memoirs* 41 (1879), v–vi; Airy to Lockyer, 3 Apr. 1871 (RGO 6/133).

51. See the letters in the RAS's eclipse correspondence, MSS 5.2. On Ranyard, see his RAS obituary.

52. Ranyard to Robert Bagnes, 7, 17, 24 Oct. 1871; Bagnes to Ranyard, 15, 22, 27 Oct. 1871; C. E. Burton to Ranyard, 6 Nov. 1871; J. Hostage to Ranyard, 2 Nov. 1871, 20 Apr. 1872, 18 Dec. 1872 (RAS MSS 5.2).

53. John Brett to Ranyard, 13 Jan. 1871 (RAS Papers 51.1).

54. *RAS Memoirs* 41 (1879), 483.

55. Ranyard to Airy, 11 Jan. 1872 (RGO 6/133).

56. Airy to Ranyard, 19 July 1873 (RGO 6/135). This policy was similar to that of other scientific publishers. Reviewers for the Royal Society specified whether pictures could be "easily gathered from the text" or should be copied with "the utmost skill and care." They also sometimes suggested how illustrations should be presented—with scales, comparison illustrations, and so forth. See Royal Society Reader's Reports, RR 9/108, 13/60, 13/217; quotation is from RR 13/63 (RS).

57. De la Rue to E. Durkin, 14 July 1873 (RGO 6/135).

58. Ranyard to Airy, 8 Feb. 1872 (RGO 6/133).

59. Biographical details are from Herbert Hall Turner, "William Henry Wesley," *RASMN* 83 (1923), 255–59; E. B. Knobel, "William Henry Wesley," *Observatory* 45 (1922), 354–55.

60. William Wesley's father was a printer and bookseller whose firm published astronomical books, imported foreign scientific journals, and served as a node in the Smithsonian international exchange network; his father-in-law,

Robert Henson, was a mineralogist whose Regent Street shop (a few blocks from the RAS) supplied the Royal Institution and British Museum. On William Wesley and Sons, see H. K. Swann (Chairman, Wheldon and Wesley, Ltd.) to author, 2 Aug. 1990 (in author's possession); on Robert Henson, see obituary for "Samuel Henson (1848–1930)," *Mineralogical magazine* 22 (1930), 395.

61. Joseph Barnard Davis, *Thesaurus craniorum: Catalogue of the skulls of the various races of man* (London: privately printed, 1867), figs. 62, 77, 86–91; St. George Mivart, "On the appendicular skeleton of the primates," *Philosophical transactions* 157 (1867), 299–420, plates 11–14.

62. Richard Owen, "Description of the cavern of Bruniquel, and its organic contents (in 2 parts)," *Philosophical transactions* 159 (1869), 517–51, woodcuts by Wesley; Richard Owen, "On the fossil mammals of Australia, Part 3, Diprotodon australis," *Philosophical transactions* 160 (1870), 519–78, plates 35, 43–50, woodcuts in text by Wesley; William Wesley to Richard Owen, 9 Dec. 1869, 25 Mar. 1886 [Owen Papers, British Museum (Natural History)].

63. Adrian Desmond, *Archetypes and ancestors: Paleontology in Victorian London, 1825–1875* (Chicago: University of Chicago Press, 1984).

64. Herbert Hall Turner, "William Henry Wesley," *RASMN* 83 (1923), 257.

65. RAS *Memoirs* 41 (1878), plates 2–8.

66. Tennant to Airy, 11 May 1872; De la Rue to Airy, 15 May 1872 (RGO 6/135). This was a situation often encountered by engravers hired to copy paintings: see Dyson, *Pictures to print*, 33, 72.

67. An example of a private observer taking important photographs was Alfred Brothers: see plate marked "Brothers—Sicily December 1870" (CAY, Box 3, File 31). On Royal Society and RAS, see Royal Society Eclipse Committee minutes, 4 Feb. 1875, 5 May 1886 (RSL); Maunder to Holden, 14 Mar. 1894 (MLS, Box 39).

68. Ranyard, "On a remarkable structure visible upon the photographs of the solar eclipse of December 12, 1871," *RASMN* 34 (June 1874), 365–69, on 366.

69. De la Rue to Airy, 15 May 1872; De la Rue to Tennant, 23 May 1872; Tennant to Airy, 27 May 1872 (RGO 6/135).

70. Ranyard, two notes passed to Airy during RAS council meeting, 28 June 1972; Airy to De la Rue, 29 June 1872; Airy to Ranyard, 29 June 1872; Ranyard to Airy, 29 June 1872; De la Rue to Airy, 1 July 1872 (RGO 6/135).

71. Ranyard to Airy, 20 June 1872; Airy to Tennant, 30 May 1872 (RGO 6/135).

72. Airy to Ranyard, 29 Jan. 1872 (RGO 6/133).

73. Quotations from De la Rue to Airy, 12 Oct. 1869 (RGO 6/122); see also Airy to Tennant, 3 Nov. 1868, 29 July 1869; Tennant to Airy, 8 Nov. 1868, 37 Mar. 1869; De la Rue to Airy, 23 Oct. 1869 (RGO 6/122). This situation parallels reproduction of works of art and book illustrations: Dyson, *Pictures to print*, 33, 38–40; Smyth, "On photographic illustration for books," *Transactions of the Scottish Society of Arts* 5 (1861), 57–92.

74. On other uses of composites in science, see Daston and Galison, "The image of objectivity," 101–3. Composite photographs were also popular in artistic photography in this period: Margaret F. Harker, *Henry Peach Robinson: Master of photographic art, 1830–1901* (London: Basil Blackwell, 1988), chaps. 3–4.

75. Ranyard, "On a remarkable structure visible upon the photographs of the solar eclipse of December 12, 1871," 366.

76. Ibid., 365–66.

77. Caveats like these were often placed on engravings to assure readers of their authenticity as scientific rather than artistic products: see De la Rue, "On heliotypography," 278–79, quotation on plate following p. 278; id., "On a photo-engraving of a lunar photograph," 171, plate following 171.

78. Ranyard, "On a remarkable structure visible upon the photographs of the solar eclipse of December 12, 1871," 365.

79. De la Rue, "On the total solar eclipse of July 18th, 1860," 338; De la Rue to Airy, 26 Aug. 1860 (RGO 6/123).

80. William Wesley, "Editorial note" (obituary of A. Cowper Ranyard), *Knowledge* (1 Feb. 1895), 25.

81. Ranyard to Airy, 20 June 1872; Lindsay to Airy, 28 June 1872 (RGO 6/135).

82. Tennant, "Report of the total eclipse of the sun, August 17–18, 1868," 33.

83. Tennant, "Report of the total eclipse of the sun, August 17–18, 1868," note by De la Rue on 48, 36. This bears an obvious resemblance to Robert Boyle's listing of witnesses of experiment, and served the same function: see Steve Shapin, "Pump and circumstance: Robert Boyle's literary technology," *Social studies of science* 14 (1984), 488–89.

84. Ranyard to Airy, 9 Sept. 1872 (RGO 6/135). On production times, see Thomas George Hill, *The Essentials of illustration: A practical guide to the reproduction of drawings and photographs for the use of scientists and others* (London: William Wesley and Son, 1915), 1; on relief etchings, Gasciogne, *How to identify prints* (cit. n. 26), sec. 6–8.

85. Michael Twyman, *Lithography 1800–1850: The techniques of drawing on stone in England and France and their application in works of topography* (Oxford: Oxford University Press, 1970), 114–18.

86. Twyman, *Lithography 1800–1850*, chaps. 1, 5; Gasciogne, *How to identify prints* (cit. n. 26), sec. 19a–e.

87. Ranyard to Airy, 9 Sept. 1872 (RGO 6/135).

88. Tennant to Lockyer, 9 Sept. 1872 (JNL, Lockyer Letters, File T).

89. De la Rue to Airy, 3 Sept. 1872 (RGO 6/135). De la Rue expressed similar opinions about mezzotint during the printing of Tennant's report: see De la Rue to Airy, 23 Oct. 1869 (RGO 6/122). Fading was a well-known problem of mezzotint: see Blanc, *Grammar of painting and engraving*, chap. 5.

90. Airy to De la Rue, 31 Aug. 1872 (RGO 6/135).

91. Ranyard to Airy, 14 Nov. and 3 December 1872 (RGO 6/135).

92. Sales figure is from advertisements for portraits, *Illustrated news of the*

world (5 Jan. 1861), 6. H. M. Hake, *Catalogue of engraved British portraits preserved in the Department of Prints and Drawings in the British Museum*, vol. 6: *Supplement and indexes* (London: British Museum, 1925), 665, and entries for individual portraits in vols. 1–4. The albums included *Drawing room portrait gallery of eminent personages* (London, J. Tallis: 1859–62), *Portraits and memoirs of the Royal Family of England* (1862), and *The statesmen of England* (1862).

93. D. J. Pound to Ranyard, 2 Dec. 1872 (RGO 6/135).

94. Ranyard to Airy, 14 Nov. and 3 Dec. 1872 (RGO 6/135).

95. Ranyard, "Estimate for each photograph copied," n.d. (RGO 6/135).

96. Ranyard to Airy, 28 Dec. 1872 (RGO 6/135).

97. Report of council, *RASMN* 35 (1875), 164; 36 (1876), 135; 37 (1877), 141; 38 (1878), 142; 39 (1879), 215. Advance copies of the first 480 pages were available in March 1878: Ranyard to Charles Young, 7 Mar. 1878 (CAY, File "Ranyard, A.C.").

98. G. L. Dupman to Airy, 10 Feb. 1874 (RGO 6/135); the article was Ranyard, "On a remarkable structure visible upon the photographs of the solar eclipse of December 12," 365–69. The pictures were also published in the *Memoirs* eclipse volume as plates 10–17.

99. Henry Draper to Ranyard, 10 Aug. 1880; Samuel Langley to Ranyard, 28 Apr. 1880; Pickering to Ranyard, 3 Aug. 1880; Ball to Ranyard 7 May, 1880 (RAS MSS 6).

100. Proctor, "The sun's corona and his spots," *Contemporary review* (Sept. 1878), 322–38.

101. George Ellery Hale, "Arthur Ranyard, 1845–1894," *Astrophysical journal* 1 (1894), 168. Ranyard tried to tempt Charles Young into writing for *Knowledge* with the promise to "do anything I could to illustrate a paper for you . . . [with] any photograph you wished reproduced": Ranyard to Young, 31 May 1893 (CAY, File "Ranyard, A.C.").

102. Huggins, "On some results of photographing the solar corona without an eclipse," repr. in William and Margaret Huggins, eds., *The scientific papers of William Huggins* (London: William Wesley and Sons, 1909), 324; Karl Hufbauer, "Lyot and the coronagraph, 1929–1939" (paper presented at the 1990 annual meeting of the History of Science Society, Seattle). Thanks to Prof. Hufbauer for a copy of his paper.

103. "The little drawing looks so beautiful: I think now we ought to have put your name to it." Turner to Wesley, 3 Mar. 1898 (RAS Papers, File "H. H. Turner"); Samuel Henson, "On a crystal of apatite," *Mineralogical magazine* 5 (1883), 198. Wesley appears to have developed a relationship with the printing firm of West, Newman and Company. Thirteen of fifteen lithographed plates in the *Monthly notices* were printed by them, at a time when almost all *Monthly notices* plates were printed by Spottiswoode and Co. Wesley also contributed short articles to *Knowledge*: see "The Formation of coral reefs" (1 Nov. 1888), 4–7; "Footprints of a prehistoric man" (1 Dec. 1888), 28–30; "The volcanoes of the Sandwich Islands"(1 Mar. 1889), 97–100.

104. Otto Boeddicker, *The Milky Way from the North Pole to 100 of south declination drawn at the Earl of Rosse's observatory at Birr Castle* (London: Longmans, Green, 1892); William Wesley and Mary A. Blagg, *IAU map of the moon: Based on the fiducial measures of S. A. Saunders and J. Franz* (London: Percy Lund, Humphries and Co., 1935).

105. Sidney Waters, "On two distribution maps of the nebulae and clusters in Dr. Dreyer's catalogue of 1888," *RASMN* 54 (1894), 526.

106. Turner to Wesley, 18 Nov. 1896 (RAS Papers, File "H. H. Turner").

107. Copeland had originally described Wesley as a "draughtsman," but changed the word to "expert": Copeland to Committee on Scientific Endowment, University of Edinburgh, 27 Jan. 1899 (ROE, File 36.262).

108. Common to Copeland, 28 Apr. 1899 (ROE, File 36.262).

109. This summary draws on Owen Gingerich, ed., *The general history of astronomy*, vol. 4: *Astrophysics and twentieth-century astronomy to 1950: Part A* (Cambridge: Cambridge University Press, 1984), part I; John Lankford, "Amateurs and astrophysics: A neglected aspect in the development of a scientific specialty," *SSS* 11 (1981), 275–303; Schaffer, "Astronomers mark time," 115–45. For the American case, see Marc Rothenberg, "Organization and control: Professionals and amateurs in American astronomy, 1899–1918," *SSS* 11 (1981), 305–25.

110. Daston and Galison, "The image of objectivity," 81–128.

111. Andrew Common, "Note on a photograph of the great nebula in Orion and some new stars near ø Orionis," *RASMN* 43 (1883), 255–57; David Gill, "The application of photography in astronomy," *Observatory* 10 (1887), 267–72, 283–94; "Astronomical photography," *Edinburgh review* 167 (1888), 23–46.

112. E. E. Barnard, "Recent stellar photography," *Sidereal messenger* 6 (1887), 58–65, esp. 62.

113. Isaac Roberts to Edward Holden, 18 Sept. 1891 (MLS, Box 49, File "Isaac Roberts, 1889–1897").

114. "Astronomical photography," *Sidereal Messenger* 7 (1888), 184.

115. Isaac Roberts to Edward Holden, 23 Sept. 1890 (MLS, Box 49, File "Isaac Roberts, 1889–1897").

116. Foster to Turner, 19 Mar. 1894; Foster to Turner, 28 Apr. 1894, copied in Solar Eclipse Committee Minutes, 2 May 1894 (RAS 54.1).

117. Turner to Foster, 28 Feb. 1894, copied in Solar Eclipse Committee Minutes, 2 May 1894 (RAS 54.1).

118. Ralph Copeland, "Total solar eclipse of January 22, 1898," *RASMN* app. to 58 (1898), 21–26; Heber D. Curtis, "The U.S. Naval Observatory eclipse expedition to Sumatra," *Proceedings of the Astronomical Society of the Pacific* 13 (1901), 205–13; John Eddy, "The Schaeberle 40-foot eclipse camera of Lick Observatory," *JHA* 2 (1971), 1–22.

119. The cameras and their use are described in Lockyer, "Eclipse notes," *Nature* (20 Apr. 1882), 577; L. Darwin, Arthur Schuster, and E. Walter Maunder, "On the total solar eclipse of August 29, 1886," *Philosophical transac-*

tions, series A, 180 (1889), 302, 311, 342, 345; H. H. Turner, "Report of the Eclipse Committee of the Royal Astronomical Society, 1889 October 3," *RASMN* 50 (1889), 2–8; W. H. M. Christie and H. H. Turner, "Report on the expedition to Sahdol," *RASMN* 58 (1898), 10, 14; Edmond Hills and Hugh F. Newall, "Report on the solar eclipse expedition to Pulgaon," *RASMN* 58 (1898), 48–50. Sir Robert Stawell Ball, *In the high heavens* (London: Isbister and Co., 1901), 84–87, 91.

120. W. H. M. Christie, H. H. Turner, and E. H. Hills, "The total solar eclipse of 1896, August 9," *RASMN* 57 (1897), 105–6; Christie and Turner, "Report of the solar eclipse expedition to Sahdol," 5–9. The instrument was a gift to the Royal Observatory of Sir Henry Thompson, a distinguished surgeon who was a major supporter of astrophysical research at Greenwich: see A. J. Meadows, *Greenwich observatory*, vol. 2: *Recent history (1836–1975)* (London: Taylor and Francis, 1975), 13.

121. Ralph Copeland to JPEC, 14 May 1898 (ROE 34.257, Letterbook 1896–1911).

122. On the design and use of coelostats, see H. H. Turner, "Some notes on the use and adjustment of the coelostat," *RASMN* 56 (1896), 408–23; H. C. Plummer, "Notes on the coelostat and siderostat," *RASMN* 65 (1905), 485–501.

123. On the place (or absence) of visual observation in eclipse plans, see Eclipse Committee minutes, 9 July 1885, 15 Apr. 1886 (RS, CMB.2, "Miscellaneous Committees 1869–1884"); Turner, "Report of the Eclipse Committee of the Royal Astronomical Society, 1889 October 3," 2–8, esp. 3–5; Andrew Common, "Report of the Joint Solar Eclipse Committee . . . for the observation of the solar eclipse of 1893 Apr. 16," *RASMN* 50 (1894), 404–8. Streamer observations are discussed in "The total solar eclipse of 1886," *Nature* (23 Sept. 1886), 497–99; Turner, "Report of the Eclipse Committee of the Royal Astronomical Society, 1889 October 3," 4; "The recent solar eclipse," *Nature* (11 May 1893), 40–42, esp. 41; "The approaching solar eclipse," *Observatory* 19 (1896), 293–98, on 294; J. Norman Lockyer, "The total solar eclipse of August 9, 1896—Report on the expedition to Kiö Island," *Philosophical Transactions,* series A, 190 (1900), 1–21, esp. 14–16; Lockyer, "What the 'Melpomenes' did at Viziadrug," *Journal of the Royal United Service Institution* (15 Aug. 1899), 861–77, esp. 864–67.

124. Eclipse Subcommittee meeting minutes, 21 July 1886 (RS CMB.2, "Miscellaneous Committees 1869–1884").

125. Lockyer to Hills, Feb. 1900 (RAS Papers, 55.2).

126. Lockyer to Copeland, 14 Apr. 1900 (ROE, 36.264, "Ralph Copeland Correspondence—Letters in 1900"). Emphasis in original.

127. Lockyer, "The total eclipse of the sun," *Nature* (17 Feb. 1898), 366.

128. John James Aubertin, *By order of the sun to Chile to see his total eclipse* (London: Kegan Paul, Trench, Trubner and Co., 1894), 141.

129. H. H. Turner to Lockyer, 4 Sept. 1898 (JNL, "Lockyer Letters," File T).

130. Ralph Copeland, "The total solar eclipse of 21 January 1898, with

some account of solar observations generally," handwritten ms. of lecture, n.d. (ROE, 36.300), 19.

131. Such feelings were not reported only by British astronomers. After his third expedition, American solar physicist C. G. Abbott wrote: "[It] would be a great pleasure to go to an eclipse and *see* it, instead of 'observing' it. . . . During totality [the writer] was merely an instrumental observer, his eyes for adjusting not for seeing, his brain for executing not for apprehending, so that he has yet to really see his first eclipse." Abbot, "A total eclipse in the South Seas," 15–16 (MLS, box "Flint Island #1," file 3, italics in original).

132. They were active in developing observatory instruments as well: Abbot and Copeland directed the expansion of the Smithsonian and Edinburgh observatories. Turner oversaw Oxford's contribution to the International Astrographic Chart, a photographic map of the heavens: see H. H. Turner, *The great star map: Being a brief general account of the international project known as the Astrographic Chart* (London, 1912).

133. For the absence of emotion during totality, see Captain William de W. Abney, "Total eclipse of the sun observed at Caroline Island, on 6th May 1883," *Philosophical transactions,* series A (1889), 119–35; Leonard Darwin, Arthur Schuster, and E. Walter Maunder, "On the total solar eclipse of August 29, 1886," *Philosophical transactions,* series A, 180 (1889), 291–350; Norman Lockyer et al., "Total eclipse of the sun, May 28, 1900," *Philosophical transactions,* series A, 198 (1902), 375–415; Hugh F. Newall, "Total solar eclipse of 1901, May 17–18," *Royal Society proceedings* 69 (1902), 209–34; Frank Dyson, "Total eclipse of the sun, 1901, May 18," ibid., 235–46; Edward Maunder, "Total eclipse of the sun, 1901, May 18," ibid., 247–61; Anne Maunder, "Preliminary note on observations of the total eclipse of the sun, 1901, May 18," ibid., 261–66. Volunteer observers who accepted positions that prevented them from seeing the eclipse were praised for their dedication to science: Lewis Swift thanked those volunteers who through "self-denial . . . lost many of the most beautiful features of the eclipse": Swift, quoted in Chicago Astronomical Society, *The solar eclipse of July 29, 1878* (Chicago: Evening Journal Book and Job Printing House, 1878), 26. See also Wallace Campbell, "A general account of the Lick Observatory–Crocker eclipse expedition to India," *PASP* 10 (1898), 127–40, esp. 138–39; Frederic Mylius, "Pictures of the eclipse," *Camera craft* (July 1900), 112–15, esp. 114.

134. This argument is inspired by Daston and Galison, "The image of objectivity," esp. 103–4, 117–23.

135. *Astronomical register* 8 (1870), 103; J. Norman Lockyer, *Contributions to solar physics* (London: Macmillan, 1874), 343–45.

136. Edward Walter Maunder, ed., *The Indian eclipse, 1898* (London: Hazell, Watson and Viney, 1899), 87–88.

137. H. Keatley Moore, in Maunder, ed., *The Indian eclipse, 1898,* 9, 91–93, 153–54; quotation on 93.

138. The first example of instructions to draw a quadrant of the corona during totality were issued in 1878 to amateurs organized by the Chicago As-

tronomical Society. They were told to "draw the outline of one quarter of the corona as accurately as possible during totality; and execute on a separate blank, after the total phase had passed, the outline of the whole corona, from memory." However, it appears that no attempt was made to produce a composite drawing based on the quadrants: Chicago Astronomical Society, *The solar eclipse of July 29, 1878,* 11. The BAA's program was imitated by other parties: see for example Alfred E. Burton, "The MIT Eclipse Expedition to Washington, Ga.," *Technology review* (July 1900), 193–213, esp. 203–5.

139. On the importance of recognizing the difference between the impact of photography in making observations, and its impact in reproduction, see Blum, *Picturing Nature,* esp. chap. 7.

140. Wilhelm Cronenberg, *Half-tone on the American basis,* trans. William Gamble (London: Country Press, 1896), 49; for opinions of dry plates in Edinburgh printers, see Charles Piazzi Smyth to Charles Young, 10 Aug. 1887 (CAY, Box 7, File 116, "Smyth, Charles Piazzi").

141. H. Jenkins, *A manual of photoengraving: Containing practical instructions for producing photoengraved plates in relief-line and half-tone* (Chicago: Inland Printer Co., 1902), 55–57.

142. Cronenberg, *Half-tone on the American basis,* 112–15; Jenkins, *A manual of photoengraving,* chap. 9.

143. Jenkins, *A manual of photoengraving,* 86–89; Alfred Walter Elson, "The Reproductive Processes of the Graphic Arts," part IV of his *Lectures on Printing* (Cambridge, Mass., 1912), 30.

144. Jussim, *Visual communication and the graphic arts* (cit. n. 2), 133, 140; see also Gasciogne, *How to identify prints* (cit. n. 26), sec. 37–39.

145. Ernest Edwards, "The art of making photo-gravures," *Anthony's photographic bulletin* 17 (1887), 430, 431.

146. The first were copies of silver positives of the Pleidaes and Andromeda made by E. E. Barnard, and were printed by the Direct Photo Engraving Co., Ltd., a London firm that produced photoengravings and photolithographs. E. E. Barnard, "On some celestial photograph made with a large portrait lens at the Lick Observatory," *RASMN* 50 (1890), 310–14, plate 3. On the Direct Photo Engraving Co., see *Kelly's directory of stationers, printers, booksellers, publishers, papermakers, &c. of England, Scotland, and Wales . . .* 13th ed. (London: Kelly's Directories Ltd., 1919), 680–82. For photomechanical reproductions of drawings, see *RASMN* 54 (1893–94), plates 6, 10; 56 (1895–96), plate 9; 58 (1897–98), plate 5; 59 (1989–99), plate 10; 60 (1899–1900), plates 3, 12.

147. Abney, "Total eclipse of the sun observed at Caroline Island, on 6th May, 1883," 119–35, quotation on 122; plates 1, 2, 10.

148. Maunder, ed., *The Indian Eclipse, 1898,* 115; Hills to Maunder, 20 Nov. 1898 (RAS 55.2); Ralph Copeland to William Wesley, 5 Apr. 1899 (ROE, 34257, Letterbook 1896–1911).

149. My thinking on the question of the "meaning" of photomechanical technology is guided by Wiebe E. Bijker, "The social construction of Bakelite:

Toward a theory of invention," in Bijker, Thomas P. Hughes, and Trevor Pinch, eds., *The social construction of technological systems* (Cambridge: MIT Press, 1987), 159–87.

150. Otto Boeddicker, *The Milky Way from the North Pole to 100 of south declination*, preface.

151. The frontispiece for the 1898 JPEC reports shows all of these flaws: *Royal Society proceedings* 64 (1898), plate 1.

152. For example, the plate accompanying the official reports of the JPEC expeditions of 1898 was "a reproduction of one of the best photographs": Edmond Hills and Hugh F. Newall. "Total solar eclipse of January 22, 1898," *RASMN* app. to 58 (1898), 52.

153. A photograph of the corona in 1900, for example, carried the caption, "The spot with cross rays on the right-hand side is a defect in the photographic plate": Christie and Dyson, "Total eclipse of the sun, 1900, May 28: Preliminary account of the observations made at Ovar, Portugal," *RASMN* 60 (1900), 392–402, quotation on plate 22. Another article from 1900 told readers, "Owing to the difficulty of reproducing the prominences and coronal extensions on the same plate, two plates have been made from negative B. . . . Plate I gives a good general idea of the photograph . . . although of course the finest details are lost, and there is less extension than on the negative." C. J. Joly et al., "The total solar eclipse of 1900," *Transactions of the Royal Irish Academy*, sec. A, 32 (1903), 276.

154. Joly et al., "The total solar eclipse of 1900," 271–98, plates 1, 2.

155. Harley Shaiken, *Work transformed: Automation and labor in the computer age* (Lexington, Mass.: Lexington Books, 1984); David Noble, *The forces of production: A social history of industrial automation* (Oxford: Oxford University Press, 1984), esp. chap. 11.

156. Carl Nemethy, "Photo-etching and printing," *American art printer* (Nov. 1891), 99–100.

157. Jenkins, *A manual of photoengraving*, 49, chap. 11; Cronenberg, *Half-tone on the American basis*, 41–44, 76, 151–54; Elson, "The reproductive processes of the graphic arts," 30; J. P. Ourdan, *The art of retouching* (New York: E. and H. T. Anthony, 1880).

158. Edwards, "The art of making photo-gravures," 430–31.

159. Cronenberg, *Half-tone on the American basis*, 77.

Chapter 5: Astrophysics and Imperialism

1. The term "black box" comes from Bruno Latour and Steve Woolgar, *Laboratory life: The construction of scientific facts* (1979; repr. Princeton: Princeton University Press, 1986).

2. Owen Hannaway, "Laboratory design and the aim of science: Andreas Libavius versus Tycho Brahe," *Isis* 77 (1986), 585–610; Steven Shapin and Simon Schaffer, *Leviathan and the air-pump: Hobbes, Boyle, and the experimental life* (Princeton: Princeton University Press, 1985), esp. chap. 7; Shapin, "The house of experiment in seventeenth-century England," *Isis* 79 (1988), 369–72.

3. This history is still in its infancy; but see Romualdus Sviedrys, "The rise of physics laboratories in Britain," *HSPS* 7 (1976), 405–36.

4. Edward Cookworthy Robins, *Technical school and college building: Being a treatise on the design and construction of applied science and art buildings* (London: Whittaker, 1887); "Research laboratory of the Royal College of Physicians, Edinburgh," *Nature* (15 Nov. 1889), 68–69; "A lab for physical and chemical research," *Nature* (1894), 217; "New Municipal Technical Institute, Belfast," *Nature* (7 Nov. 1907), 234–35; Charles Baskerville, "Laboratory organization," *Science* (1 May 1908), 681–86; Baskerville, "Some principles in laboratory construction," *Science* (13 Nov. 1908), 665–76; Augustus Hill, "Suggestions for the construction of chemistry labs," *Science* (22 Oct. 1909); Robert Millikan, "Enlarged facilities for physical research," *University of Chicago magazine* (July 1911), 17–18; "The new Ryerson Laboratory," *University of Chicago Magazine* (July 1913), 41–43; Albert Carman, "The design of a physical laboratory," *Brickbuilder* (Dec. 1911), 257–60; Paul Forman, John Heilbron, and Spencer Weart, "Physics circa 1900: Personnel, finding, and productivity of the academic establishments," *HSPS* 5 (1975), 109–13.

5. "New buildings of the University of Liverpool," *Nature* (17 Nov. 1904), 63–65. Holt was a long-time patron of physics at Liverpool: see Bruce Hunt, "Experimenting on the ether: Oliver Lodge and the great whirling machine," *HSPS* 16 (1986), 111–34.

6. "New physics laboratory of Owens College," *Nature* 58 (27 Oct. 1898), 621–22.

7. Charles Baskerville, "Some principles in laboratory construction," 665–76, quotation on 669.

8. Schuster, quoted in "New physics laboratory of Owens College," *Nature* (27 Oct. 1898), 622.

9. This discussion follows arguments developed in Mari E. W. Williams, "Astronomical observatories as practical space: The case of Pulkowa," in Frank A. J. L. James, ed., *The development of the laboratory: Essays on the place of experiment in industrial civilization* (New York: American Institute of Physics, 1989), 118–36.

10. Julius Scheiner, *A Treatise on astronomical spectroscopy*, trans. and ed. Edwin B. Frost (Boston: Ginn and Co., 1894), 41–42, 70.

11. Hale to Campbell, 5 Apr. 1897 (MLS, Box 22, File "G. E. Hale"), 5; on insulating spectroscopes, Campbell to Hale, 19 Apr. 1897 (MLS, Copybook Y), 26–27; Frost to Young, 11 Nov. 1900 (CAY, Box 6).

12. On clock drives, see E. E. Barnard, "Recent stellar photography," *Sidereal messenger* 6 (1887), 58–65, esp. 63; 1884 RAS Gold Medal Address, *RASMN* 44 (1884), 221–23; Common, "Note on a method of giving long exposures in astronomical photography," *RASMN* 45 (1884), 25–27; Common, "Photography as an aid to astronomy," *Royal Institute proceedings* 11 (1886), 367–75; Charles Young, "Astronomical photography," *New Princeton review*, series 5:3 (1887), 354–69.

13. Insulation and temperature controls are described in Williams, "Astro-

nomical observatories as practical space," 118–36, esp. 125, 131; George Forbes, "The Royal Observatory, Greenwich." *Good words* 13 (1872), 792–96 and 855–58, esp. 794–95, 856; telescope mounts and drives are discussed in Alfred Van Helden, "Telescope building, 1850–1900," in Owen Gingerich, ed., *The general history of astronomy*, vol. 4: *Astrophysics and twentieth-century astronomy to 1950: Part A* (Cambridge: Cambridge University Press, 1984), 40–58, esp. 44.

14. William and Margaret Huggins, eds., *The scientific papers of Sir William Huggins* (London: Wesley and Son, 1909), 8.

15. Simon Schaffer, "Where experiments end: Tabletop trials in Victorian astronomy," in Jed Buchwald, ed., *Scientific practice: Theories and stories of doing physics* (Chicago: University of Chicago Press, 1995), 257–99.

16. Daniel Headrick, *The tools of empire: Technology and European imperialism in the nineteenth century* (Oxford: Oxford University Press, 1981); *The tentacles of progress: Technology transfer in the age of imperialism, 1850–1940* (Oxford: Oxford University Press, 1988).

17. Alfred Crosby, *Ecological imperialism: The biological expansion of Europe, 900–1900* (Cambridge: Cambridge University Press, 1986).

18. T. J. Barron, "Science and the nineteenth-century Ceylon coffee planters," *Journal of imperial and commonwealth history* 16 (1987), 5–21. Barron also notes that some of the most important coffee-processing machinery was designed by former railroad engineer turned farmer John Brown.

19. George C. Allen and Audrey G. Donnithorne, *Western enterprise in Indonesia and Malaya: A study in economic development* (London: George Allen and Unwin, 1957), 227.

20. David H. Breen, *The Canadian prairie West and the ranching frontier, 1874–1924* (Toronto: University of Toronto Press, 1983), quotation on 23. See also Jimmy Skaggs, *Prime cut: Livestock raising and meatpacking in the United States, 1607–1983* (College Station: Texas A&M Press, 1986).

21. Jonathan C. Brown, *A socioeconomic history of Argentina, 1776–1860* (Cambridge: Cambridge University Press, 1979), 223–27; Peter H. Smith, *Politics and beef in Argentina: Patterns of conflict and change* (New York: Columbia University Press, 1969), 33–34.

22. "Indian railways," *Cornhill magazine* 20 (1869), 68–80, quotation on 79.

23. William Cronon, *Nature's metropolis: Chicago and the great West* (New York: Norton, 1991), 59.

24. Robert H. Mattoon, Jr., "Railroads, coffee, and the growth of big business in São Paolo, Brazil," *Hispanic American historical review* 57 (1975), 273–95; Warren Dean, *With broadax and firebrand: The destruction of the Brazilian Atlantic Forest* (Berkeley: University of California Press, 1995), quotations on 185, 181.

25. A good review is William Beinart, "Empire, hunting, and ecological change in Southern and Central Africa," *Past and present* 128 (1990), 162–86. On economic development and transitions in hunting, E. I. Steinhart, "Hunters, poachers, and gamekeepers: Towards a social history of hunting in colonial

Kenya," *Journal of African history* 30 (1989), 247–64. On the ritual of big-game hunting and safari as "invented tradition," see William Storey, "Big cats and imperialism," *Journal of world history* 2 (1991), 135–73, quotation on 138. Other efforts to fit British rule into "native" forms are described in Eric Hobsbawn and Terence Ranger, eds., *The invention of tradition* (Cambridge: Cambridge University Press, 1987).

26. This summary draws heavily on Ramachandra Guha and Madhav Gadgil, "State forestry and social conflict in British India," *Past and present* 123 (1989), 141–77, quotation on 169; see also E. A. Smithies, *India's forest wealth* (Oxford: Oxford University Press, 1925).

27. Mattoon, "Railroads, coffee, and the growth of big business in São Paolo, Brazil," quotation on 292.

28. Captain Edward Davidson, R. E., *The railways of India* (London: Spon, 1868), 109.

29. George C. Allen and Audrey G. Donnithorne, *Western enterprise in Indonesia and Malaya: A study in economic development* (London: George Allen and Unwin, 1957), 227.

30. "Indian railway reform," *Westminster review* 36 (1869), 1–36, quotation on 28.

31. Captain Edward Davidson, R.E., *The railways of India: With an account of their rise, progress, and construction, written with the aid of the records of the India Office* (London: Spon, 1868), quotation on 3.

32. Rowland MacDonald Stephenson, *Railways: An introductory sketch, with suggestions in reference to their extension to British colonies* (London: John Weale, 1850), quotation on 7, 46–48.

33. Charles Pritchard, "A journey in the service of science: Being a description of the phenomena of a total solar eclipse, chiefly as observed at Gujuli, Spain, on July 18, 1860," (part 2) *Good words* (Oct. 1860), 694–701, quotations on 695.

34. Tennant to secretary, Government of India, 12 June 1868 (RGO 6/122).

35. JPEC planning is described in Lockyer, "The approaching total eclipse of the sun," *Nature* (17 June 1897), 156; Copeland quotation, "The total solar eclipse of 21st January 1898," handwritten ms., n.d., 2 (ROE, 36.300); for stations described in relation to their place on railroad lines, see "Stations for observing the total eclipse of the sun in January 1898," *Nature* (2 Sept. 1897), 424–25; "The coming total solar eclipse," *Nature* (2 Dec. 1897), 105; "Arrival of the eclipse parties at Bombay," *Nature* (6 Jan. 1898), 230; William Christie and Herbert Hall Turner, "Report of the expedition to Sahdol," *RASMN* app. to 58 (1899), 2. When building the Royal Observatory at Edinburgh in the 1890s, contractors first laid a railroad track to Observatory Hill: Hermann A. Brück, *The story of astronomy in Edinburgh* (Edinburgh: Edinburgh University Press, 1983), 55.

36. Sidney G. Burrard to Christie, 14 Sept. 1897 (RGO 7/195).

37. James F. Tennant, "Report of the total eclipse of the sun, August 17–18, 1868," *RAS memoirs* 37 (1869), 6–7.

38. Airy to H. Meriwale, 31 May 1871 (RGO 6/134).

39. Trevelyn, quoted in Michael Adas, *Machines as the measure of men: Science, technology, and ideologies of Western dominance* (Ithaca: Cornell University Press, 1989), 228.

40. S. G. Burrard to William Christie, 23 Sept. 1897 (RGO 7/195).

41. L. A. Lugard to H. A. Hugard, 30 Sept. 1897 (ROE 36.262).

42. William Pole, "The eclipse expedition to Spain," *Macmillans* (Sept. 1860), 406–16, quotation on 410; see also K. H. Vignoles, *Charles Blacker Vignoles: Romantic engineer* (Cambridge: Cambridge University Press, 1982), 161.

43. Charles Strahan to William Christie, 17 Feb. 1897 (RGO 7/195).

44. S. G. Burrard to William Christie, 28 Dec. 1897 (RGO 7/195).

45. W. H. R. Rivers, "Report on anthropological research outside America," in *Reports upon the present condition and future need of the science of anthropology* (Washington: Carnegie Institution of Washington, Publication no. 200), 5–28, quotation on 7.

46. Bruno Latour makes a similar point in "Give me a laboratory and I will raise the world," in Karin Knorr-Cetina and Michael Mulkay, eds., *Science observed: Perspectives on the social study of science* (London: Sage, 1984), 141–70.

47. Charles Pritchard, "Historical sketches of solar eclipses," *Good words* (1871), 637.

Index

In this index an "f" after a number indicates a separate reference on the next page, and an "ff" indicates separate references on the next two pages. A continuous discussion over two or more pages is indicated by a span of page numbers, e.g., "57–59." "Passim" is used for a cluster of references in close but not consecutive sequence.

Printed in the USA
CPSIA information can be obtained
at www.ICGtesting.com
JSHW021436221024
72172JS00002B/24